William Jackson Hooker

A Second Century of Ferns

William Jackson Hooker

A Second Century of Ferns

ISBN/EAN: 9783742844958

Manufactured in Europe, USA, Canada, Australia, Japa

Cover: Foto ©berggeist007 / pixelio.de

Manufactured and distributed by brebook publishing software
(www.brebook.com)

William Jackson Hooker

A Second Century of Ferns

A SECOND

CENTURY OF FERNS;

BEING

FIGURES WITH BRIEF DESCRIPTIONS

OF

One Hundred

NEW, OR RARE, OR IMPERFECTLY KNOWN SPECIES OF

FERNS;

FROM VARIOUS PARTS OF THE WORLD:

BY

SIR WILLIAM JACKSON HOOKER, K.H.,

LL.D., F.R.A. AND L.S., &c. &c.

DIRECTOR OF THE ROYAL BOTANICAL GARDENS, KEW.

LONDON:

DULAU & CO., 37, SOHO SQUARE.

———

MDCCCLXIV.

TO

DR. GEORGE METTENIUS,

PROFESSOR OF BOTANY, AND DIRECTOR OF THE BOTANIC GARDEN
AT THE UNIVERSITY OF LEIPZIG.

THE ABLE AUTHOR OF

"FILICES HORTI BOTANICI LIPSIENSIS,"

AND OF VARIOUS MEMOIRS,

"UBER EINIGE FARRNGATTUNGEN,"

THIS CENTURY OF FERNS

IS DEDICATED,

IN TESTIMONY OF GREAT RESPECT AND ESTEEM, BY

THE AUTHOR.

Royal Gardens, Kew.
MAY, 1861.

PREFATORY NOTICE.

Of all the Families of plants, perhaps there is none that needs so much to be illustrated by figures as the Ferns. They are, from their variableness of character, and owing to the different forms they exhibit in different individuals of the same species, and even in different parts of the same individual, especially in the very compound kinds, exceedingly difficult of verbal definition, and hence the descriptions of writers have been so greatly misunderstood, even those that have been the most full and most accurately worded.

It was the want of such Fern-figures, as helps to a more thorough knowledge of these lovely plants, that induced us to devote the whole of the last volume (the Tenth) of our " Icones Plantarum," exclusively, to this Family, and we think Mr. Pamplin did right in issuing copies of this volume separate, to those who might not care to possess the entire work, and under the title of a "Century of Ferns." So favourably was this volume received that it has induced the Author to publish another and "Second Century," the volume now before us.

There is indeed no scarcity of materials for many such volumes, could sufficient patronage be insured ; for the discovery of new species has of late years been quite extraordinary, in different parts of the world; and it would be still more extraordinary if we could persuade ourselves

that all that go by the name of *new* discoveries, were really
so. If we take Mr. Moore's useful and laborious "Index
Filicum" as a test of the numerical statistics of Ferns, the
number appears great indeed. The eleventh part of this
"Index" has recently been issued, bringing the Catalogue
down to the letter C, and to the end of the genus *Cuspidaria*.
Thus the work at present only embraces the Genera (alpha-
betically arranged) of the first three letters of the alphabet
(A—C). These include 47 Genera and 1069 species! The
total of such genera to be recorded are 186, of which about
one-eighth only have been elaborated, so that if we consider
the remaining seven-eighths of the genera to possess the
same proportional amount of species, which may possibly be
the case, this brings us to 8000 different kinds of Ferns
described in books by persons worthy of credit from their
name and character; to say nothing of the multitudes of
synonyms, which are legion.

But it must be recollected that Mr. Moore necessarily
reckons, as *species* of authors, a great number which he has
no means of verifying, nor of judging whether they ought, or
ought not to be rejected, as he has done in the case of many
others:—and of which there is too much reason to fear that a
large proportion would come into the latter category. If we
reckon the number of well-ascertained Ferns at half the
amount enumerated in the "Index Filicum," viz. at 4000
species, it will perhaps nearly accord with the truth.

Nothing will so much tend to a correct knowledge of these
as accurate figures, published on a cheap scale, as are the two
"Centuries" now under consideration:—and we have had

the privilege of bringing into notice many novelties of great
interest derived from the researches of our friends and corres-
pondents in different quarters of the globe, from some countries of which, till now, the vegetable productions were
comparatively unknown to the man of science. We allude
particularly to the Fiji Islands, to Japan and China, to
Borneo, Tropical Africa, East and West, &c. &c. ; the results
of the labours of Brackenridge, Milne, Macgillivray, Harvey,
Seemann (Fiji); Wright, Wilford, Urquhart, Alcock, Hodgson (China and Japan); Low, Motley, Lobb (Borneo);
Vogel, Baikie and Barter, Kirk, in Livingstone's Expedition
(Tropical Africa), and others too numerous to mention, but
whose names stand recorded in these pages. Not a few we
trust will be found faithfully and unmistakeably exhibited in
our two Centuries; and many, for which we have no space
here, will be received, and if possible figured in our " Species
Filicum,*" now in course of Publication.

Royal Gardens, Kew,
May, 1861.

* "Species Filicum; being descriptions of the known Ferns, particularly of such as exist
in the Author's Herbarium, or are, with sufficient accuracy, described in works to which he
has had access; accompanied with numerous figures, by Sir W. J. Hooker." Of this work
three volumes have appeared, with 210 plates. Vol. IV. is now in the press.

INDEX.

TAB. I.

Trichomanes Henzaianum, *Parish.*

Caudice filiformi repente ramoso parce nigro-tomentoso, frondibus parvis remotiusculis brevi-stipitatis obovato - subflabelliformibus membranaceis subnitidis læte viridibus vix semiunciam longis marginibus magis minusve irregulariter lobatis vix pinnatifidis lobis brevibus obtusis, venis apice liberis primariis paucis subflabellato-pinnatis satis distinctis, secundariis iis parallelis arctis deliculatis venulis transversis junctis et ita frondibus minute reticulatis, involucris in lobis frondium venas primarias terminantibus omnino intramarginalibus textura frondis infundibuliformibus, limbo dilatato integro, stipite gracili vix lineam longo.

Trichomanes Henzaianum, *Parish in litt.*

HAB. Detected by *Mr. Henzai,* and the *Rev. C. S. P. Parish,* partially clothing the trunks of trees at Moulmein, 1859.

Dr. Van Den Bosch, of Leyden, distinguished by his writings and admirable figures of Mosses, is now happily engaged on a monograph, with numerous figures on the beautiful family, among Ferns, of *Hymenophyllaceæ.* He has already given to the world a valuable "Synopsis" of the group, with full descriptions of new or critical species. Our present individual, will rank near to *Tr. sublimbatum* of Mueller, to which Van Den Bosch refers, and probably quite correctly, my Java form of *Tr. muscoides,* (Sp. Fil. i. p. 117). From that, however, our plant may be known by its much smaller size, and different form, greener color, more delicate texture; but, above all, by the involucre, of which the limb in *T. sublimbatum* extends to the margin of the frond, while in our plant, the lobe of the frond extends much beyond the involucre; and has, indeed, the appearance of half an involucre attached to the frond, as occurs in not a few davalliaceous plants.

Fig. 1. Plants, *nat. size,* (from a drawing by Mr. Parish). *f.* 2, 3. Portions of plants *slightly magnified. f.* 4. Single fertile frond. *f.* 5. Two involucres, one laid open showing the sorus. *f.* 6. Portion of the frond, to show the cellular structure. *f.* 7. Receptacle and capsules. *f.* 8. Capsule; *more highly magnified.*

CENT. 2. T. 1.

Tab. 1.

TAB. II.

WOODSIA (HYMENOCYSTIS) POLYSTICHIOIDES, *Eaton.*

Spithamæa ad pedalem dense cæspitosa, caudice subnullo, frondibus subcoriaceo-membranaceis opacis lanceolatis pinnatis, pinnis patentibus numerosis approximatis sessilibus 6-7 lineas longis lanceolatis obtusis basi cuneato-truncatis sursum acute auriculatis junioribus sparsim paleaceis villosisque demum glabris margine integerrimis vel apicem versus obsoletissime crenatis, costa indistincta, venis immersis simplicibus vel furcatis liberis ad marginem apice soriferis, involucro e squamis 4-5 tenui-membranceis in orbem dispositis imbricatis longe ciliatis, stipitibus castaneis rachique straminea nitidissimis deciduo paleaceis.

Woodsia (Hymenocystis) polystichoides, *Eaton in Wright's Herb. of Ringgold and Rodgers U. S. North Pacif. Explor. Exped.*

HAB. Hakodadi, Japan, *C. Wright.*

A very remarkable and very pretty Fern, for which I am indebted to Mr. Wright and Mr. Eaton, who observe (in litt.) that this ought, perhaps, to be made the type of a new genus; for that "the parts of the indusium imbricate over each other." It is however, I think, very difficult, where the indusium (or involucre) is of so very delicate and fragile a nature, to say whether it is of the structure now mentioned, or whether, being first entire, it may not afterwards burst from the top into a few unequal valves, which may appear to be imbricated, and as is the case in the group or subgenus *Physematium*, Klfs. (Hymenocystis, *C. A. Meyer*), to which Mr. Eaton has, as it appears to me, properly referred it.

Fig. 1. Fertile pinna, seen from beneath. *f.* 2. Portion of the same. *f.* 3. Involucre partially closed. *f.* 4. Involucre open, and showing the sorus. *f.* 5. Portion of a valve of the involucre. *f.* 6. Capsule :—*magnified.*

Tab. II

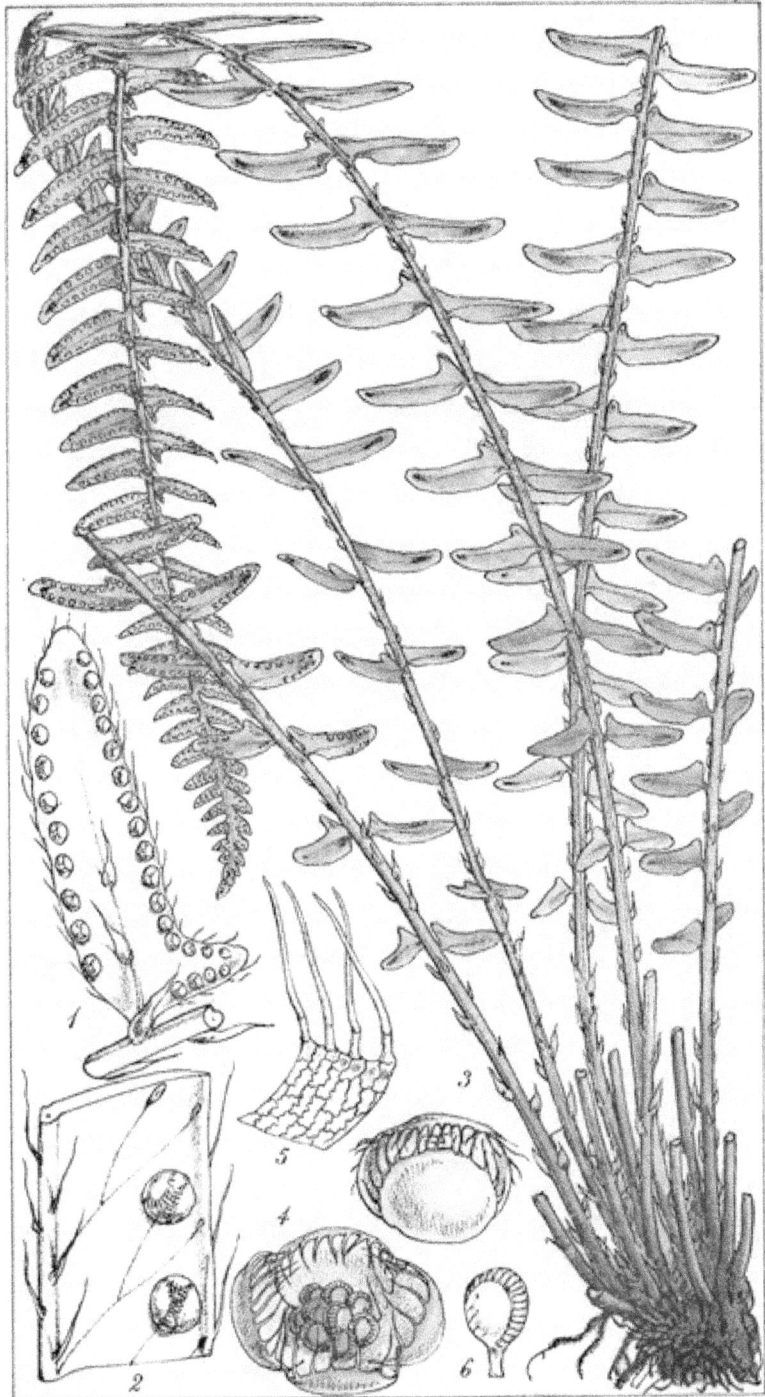

TAB. III.

ASPLENIUM (EUASSPLENIUM) LUGUBRE, *Hook.*

Glabrum, colore toto nigricante, caudice repente crassiusculo radiculoso, frondibus cæspitosis brevi-stipitatis
spithamæis ad pedalem lato-lanceolatis inferne attenuatis
pinnatis apicem versus pinnatifidis, pinnis sessilibus horizontaliter patentibus segmentisque lanceolato-falcatis vix
acuminatis inæqualiter subduplicato-serratis membranaceis
rigidis subopacis, venis simplicibus v. furcatis apicibus
intra marginem clavatis, soris versus apicem pinnarum vel
segmentorum, involucris angustis nigris, stipitibus rachibusque villis paleaceis aterrimis patentibus crinitis.

HAB. Kina Ballu, Borneo, *Hugh Low, junr. Esq.*

A very peculiar looking *Asplenium*, entirely of a black
colour in its dried state, having the short stipes and rachis
clothed with patent, long-spreading, intensely black, paleaceous flexuose hairs, or scales. The appearance of the entire
plant is that of having grown in water, and the pinnæ and
segments are more or less erose, and jagged at the margin,
and the substance is formed of closely compacted cells, in the
younger and subpellucid specimens exhibiting a minutely
reticulated appearance, when held between the eye and the
light.

Fig. 1. Portion of a pinna to show the venation. *f.* 2. Portion of a fertile pinna. *f.* 3. Scale from the rachis:—*magnified.*

CENT. 2, T. 3.

Tab. III

TAB. IV.

STRUTHIOPTERIS ORIENTALIS, *Hook.*

Elata, frondibus ovatis ovato-oblongisve pinnatis, pinnis
sterilibus pinnatifidis submembranceis laciniis ovatis obtusis,
fertilibus lato-linearibus coriaceis planis, involucris arcte
appressis dorsum totum tegentibus intense badiis nitidissimis integerrimis, demum ætate patentissimis erosis, stipite
rachi costisque inferne deciduo-paleaceis.
Struthiopteris Germanica, *Eaton in Wright's Herb. of U. S.
N. Pacif. Expl. Expd. of Ringgold and Rodgers.*

HAB. Sikkim Himalaya, elev. 12,000 ft., *Drs. Hooker and
Thomson.* Assam, *Simons, in Herb. Lady Lyell.* Hakodadi, Japan, *C. Wright.*

A single glance at the fertile pinnæ of this fine species,
is sufficient to assure any one of its distinctness from *S.
Germanica,* of Willdenow (*Pensylvanica* of the U. States
Botanists); not only are they much longer and broader and
flatter, (less cylindrical) and never moniliform; but the involucre is of a very different nature, so broad as completely
to cover the back of the pinnules, the entire edges meeting
at the back, and never breaking up into uniform segments;
and the texture is thin and membranaceous, but firm, very
glossy, and of a very dark chestnut colour, suddenly con-
duplicate, and pressed close to the sori on the back; whereas
in *S. Germanica,* the moniliform fertile pinnæ have the involucre rolled back as it were, so as to cover the sori, and of
the same texture and color as the pinna itself. The sterile
frond too, which is much attenuated at the base, is here
abrupt. As a species, it may probably be found to have as
extensive a range in the Eastern, as *S. Germanica* has in the
Western world, (including Europe in this region), for it has
already been found in Sikkim Himalaya. I detected one
specimen in a collection of Assam Ferns; and it appears
again in the Northern Island of Japan, Hakodadi.

Fig. 1. Small fertile pinna. *f.* 2. Portion of a fertile
pinna; *magnified. f.* 3. Portion of a sterile frond; *nat.
size. f.* 5. Portion of a sterile pinna, showing the venation;
magnified.

Tab IV.

Fitch. del. et lith.

Pamplin, imp.

TAB. V.

GRAMMITIS (CALYMMODON) CLAVIFER, *Hook.*

Caudice crassiusculo repente, frondibus vix stipitatis digitalibus firmis rigidis dense cæspitosis lineari-lanceolatis pinnatis, pinnis remotis patentibus apice piliferis, *sterilibus*
angustissime linearibus, *fertilibus* spathulatis acutis margine
superiore reflexo, vena (seu costa) solitaria infra apicem
terminante clavata sorifera, soris solitariis oblongis apicem
dilatatum pinnæ occupantibus, rachi angusta lineari-suba-
lata patentim villosa.

HAB. Kina Ballu, Borneo, *Hugh Low, junr. Esq.*

As far as I can judge from Fée's figure and description of
his *Plectopteris gracilis*, it is identical with the *Calymmodon
cucullata* of Presl, and the *Grammitis cucullata* of Blume, and
of J. Smith; and no less so with *Grammitis denticulata* of
Blume, (*Polypodium cucullatum* of Nees and Blume in Nov.
Act. Acad. Nat. Cur., and *P. denticulatum of Bl. Syn. Fil.
Jav.*) If this be correct, this genus (or section of *Polypodium*
or *Grammitis*, according to the views of Botanists,) is reduced
to a solitary species; but to this I have added the very
elegant Fern now before us, which I prefer retaining in
Grammitis. It is readily distinguished by the more slender
habit, and deeply and narrowly divided frond, rather pinnate
(with the rachis narrow-winged) than pinnatifid; the sterile
pinnæ are narrow linear almost acicular, while the fertile ones
are truly spathulate or claviform, and mostly confined to the
middle or upper part of the frond.

I may observe that the *Grammitis cucullata*, of an un-
usually large size, is also found on Kina Ballu, by Mr. Hugh
Low, junr.

Fig. 1. Portion of a fertile frond, seen from beneath. *f.* 2.
Single fertile pinna. *f.* 3. Capsule :—*magnified.*

CENT. 2. T. 5.

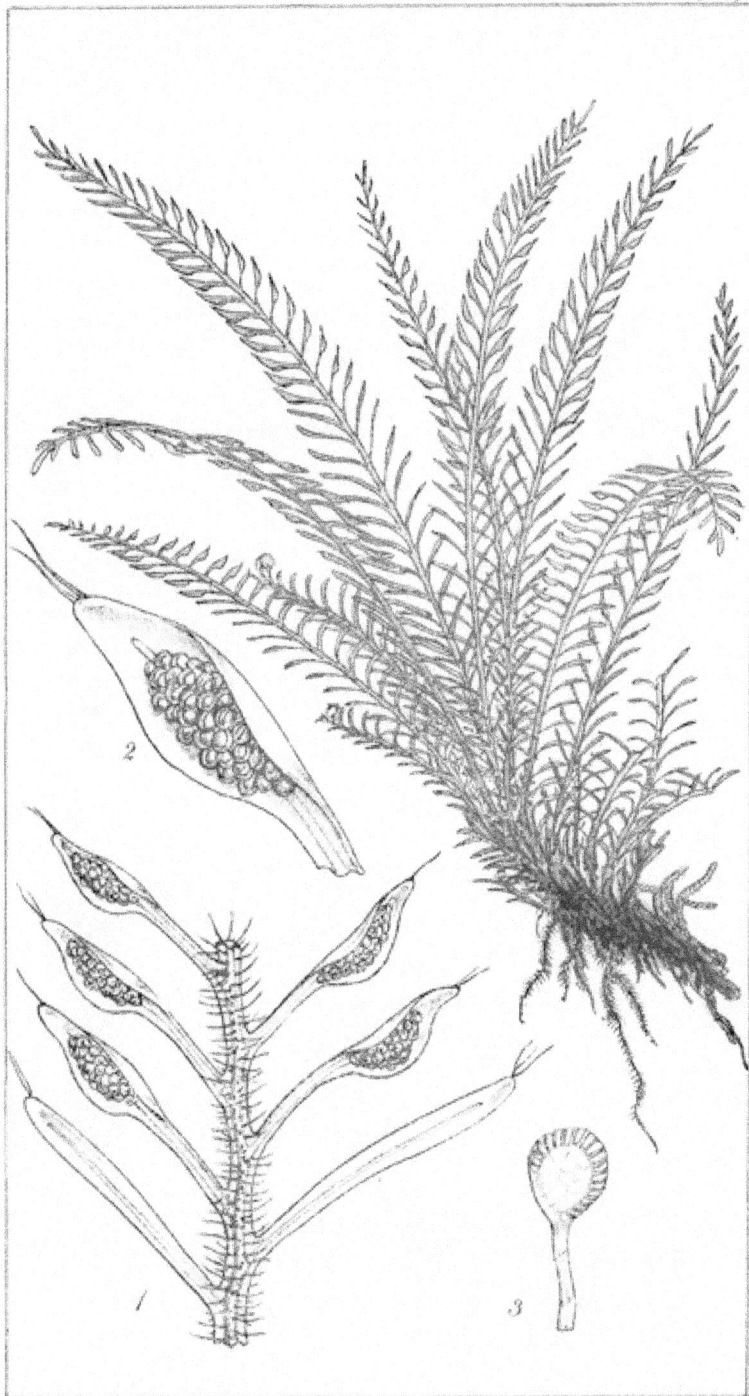

Tab. V.

TAB. VI.

POLYPODIUM ANDINUM, *Hook.*

Caudice brevi repente, frondibus membranaceis cæspitosis digitalibus ad semipedalem oblongo-linearibus vix acuminatis ad basin attenuatis fulvo-villosis ciliatisque ad marginem solummodo pinnatifidis, lobis brevibus obtusis, costa tenui, venis furcatis intra marginem desinentibus, venula superiore perbrevi apice sorifera, soris globosis subellipticisve prope costam utrinque uniserialibus singulo lobo oppositis.

HAB. Andes of Quito, on the banks of the river Hondacha, *Jameson, n.* 780. On Mount Picóte, near Moyobambu; Peru, *C. W. Nilson (in Spruce's Plants of Peru, n.* 4780*).

A very pretty and very peculiar species of the extensive Genus *Polypodium,* of which I do not find any description, and which seems confined to the Andes of Ecuador and Peru; at least, I have seen it from no other quarter. It is remarkable in the almost ligulate form of the small thin and membranaceous fronds, cut at the margin, with great regularity, into very short and obtuse lobes; the whole, on both surfaces, and at the margin, clothed with long, but rather sparse fulvous hairs. The color is pale green; in the older specimens stained with yellow and brown. The fronds seem destitute of stipes, and are decurrent to their very base, where the costa is often blackish. It may rank near the West Indian *P. Serricula* of Fée, but that has much narrower fronds, deeply pinnatifid, the lobes 1-veined, and the sorus placed within the lobe.

Fig. 1. Portion of the sterile frond, showing the venation. *f.* 2. Portion of a fertile frond :—*magnified.*

Tab.XI.

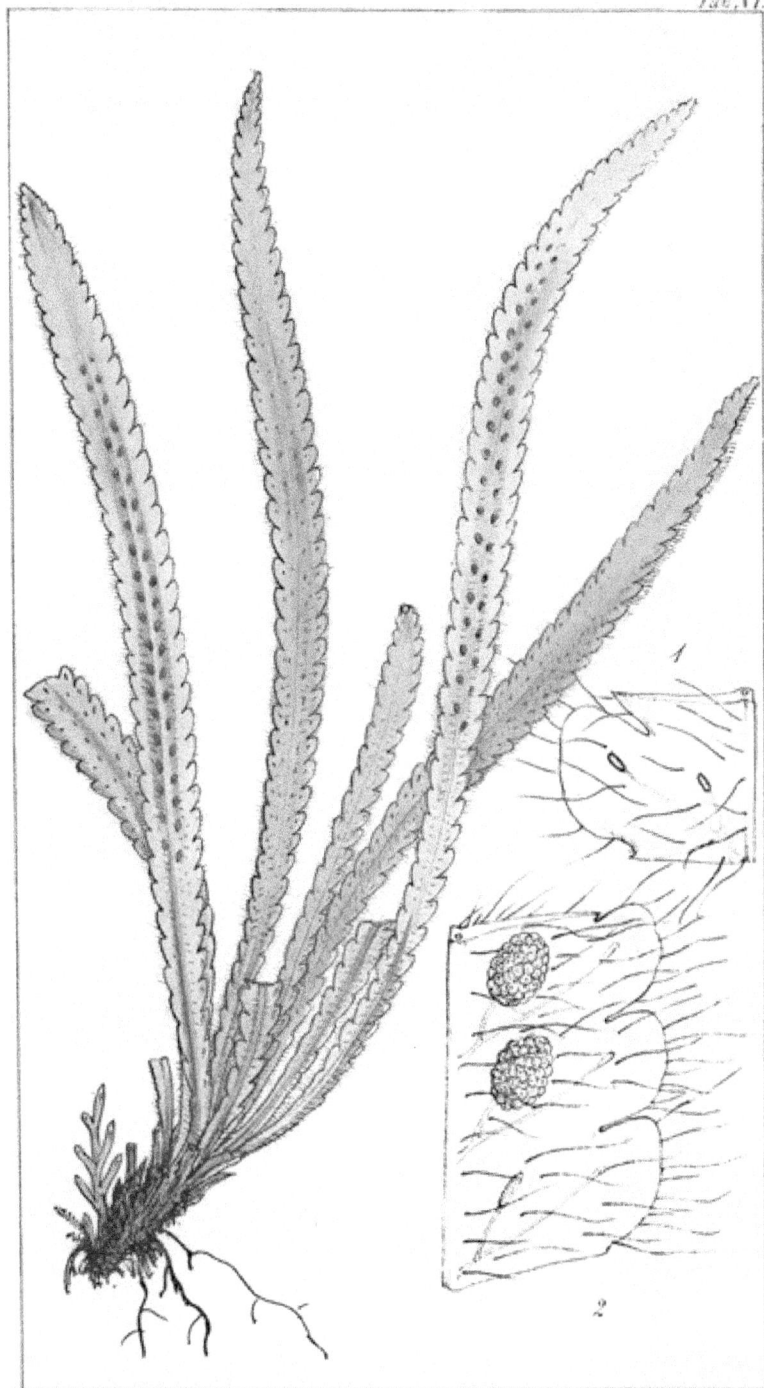

TAB. VII.

GRAMMITIS CORDATA, *Sw.—var.* subbipinnata.

Caudice brevi crasso copiose radiculoso superne paleaceo, stipitibus cæspitosis 1-2-uncialibus rachique deciduo squamosis intense nigro-ebeneis nitidissimis, frondibus erectis flexuosis curvatisve subcoriaceis 3-4-uncialibus ad spithamæam supra viridibus nudis subtus dense imbricatis ferrugineo-palcaceis pinnatis, squamis ovato-lanceolatis magis minusve longis acuminatis subciliato-dentatis subintegerrimisque, pinnis semiunciam ad 1¼ unciam longis remotinsculis sessilibus cordato-oblongis oblongisve horizontaliter patentibus integris lobato-pinnatifidis magis minusve profundis non raro iterum pinnatis rarius subauriculatis, venis liberis furcatis apice clavatis, soris oblongis.

Grammitis cordata, *Sw. Syn. Fil. p.* 23 *and* 217. *Willd. Sp, Pl.* 5. *p.* 142. Gymnogramme cordata, *Schlect. Adumbr. Pl. p.* 16. *Hook. et Grev. Ic. Fil. t.* 156. Acrostichum cordatum, *Th. Fl. Cap. p.* 732.

HAB. S. Africa; throughout the Cape Colony, I believe, plentiful, extending eastward to Uitenhage, and the elevated mountains of Macalisberg (*Ecklon and Burke*). St. Helena, elev. above the sea, 2400 feet. *Dr. Alexander, R. N.*; in Herb. Nostr., and *Mr. Houghton,* Herb., Trin. Coll. Dubl., et Nostr.

Kunze, and following him, all succeeding authors have pronounced that the admirable figure of *Gymnogramme "cordata,"* of Dr. Greville in Ic. Fil., is not the *Grammitis cordata* of Swartz; but they refer it to the *G. Capensis* of Sprengel. We maintain, that it perfectly accords with all the essential characters and full descriptions of the illustrious Swede : but it ill accords with what Kunze figures and describes as the *G. Capensis ;* nevertheless, we are quite willing to declare our opinion that the two are varieties of each other, for we can trace them through their several stages in our Herbarium, in the following forms or varieties. 1. *Pinnata ;* pinnis oblongis subintegerrimis. Ceterach Capensis, *Kze. in Analect Pterid. p.* 13. *t.* 8. *Fée, Gen. Fil. Tab.* 30. *f.* 4. *(one pinna slightly pinnatifid).* 2. *Pinnato-pinnatifia ;* pinnis cordatis profunde pinnatifidis. Grammitis cordata, *Sw. l. c. and* Gymnogramme cordata, *Hook. et Grev. l. c.* 3. *Subbipinnata ;* pinnis angustooblongis profunde pinnatifidis pinnatisque. Gymnogramme Capensis, *Spr. in Zeyh. Pl. Cap. (Herb. Nostr.) Kze. in Linnæa,* 6. *p.* 183. Ceterach Capensis, *Fée, Gen. Fil. Tab.* 30. *f.* 3. *(et Tab. Nostr.* VII*).* I may add, a 4th state, or a subvariety of the latter. 4. *Nudiuscula ;* frondibus parce palaceis, squamis minoribus subintegerrimis. Gymnogramme Namaquensis, *Pappe and Rawson, Syn. Fil. Afr. Austr. p.* 42. This is found both at the Cape, and in St. Helena.

Fig. 1. Scale of the frond. *f.* 2. Pinnule. *f.* 3. Pinnule showing the venation, and two sori :—*magnified.*

CENT. 2. T. 7.

Tab. VII.

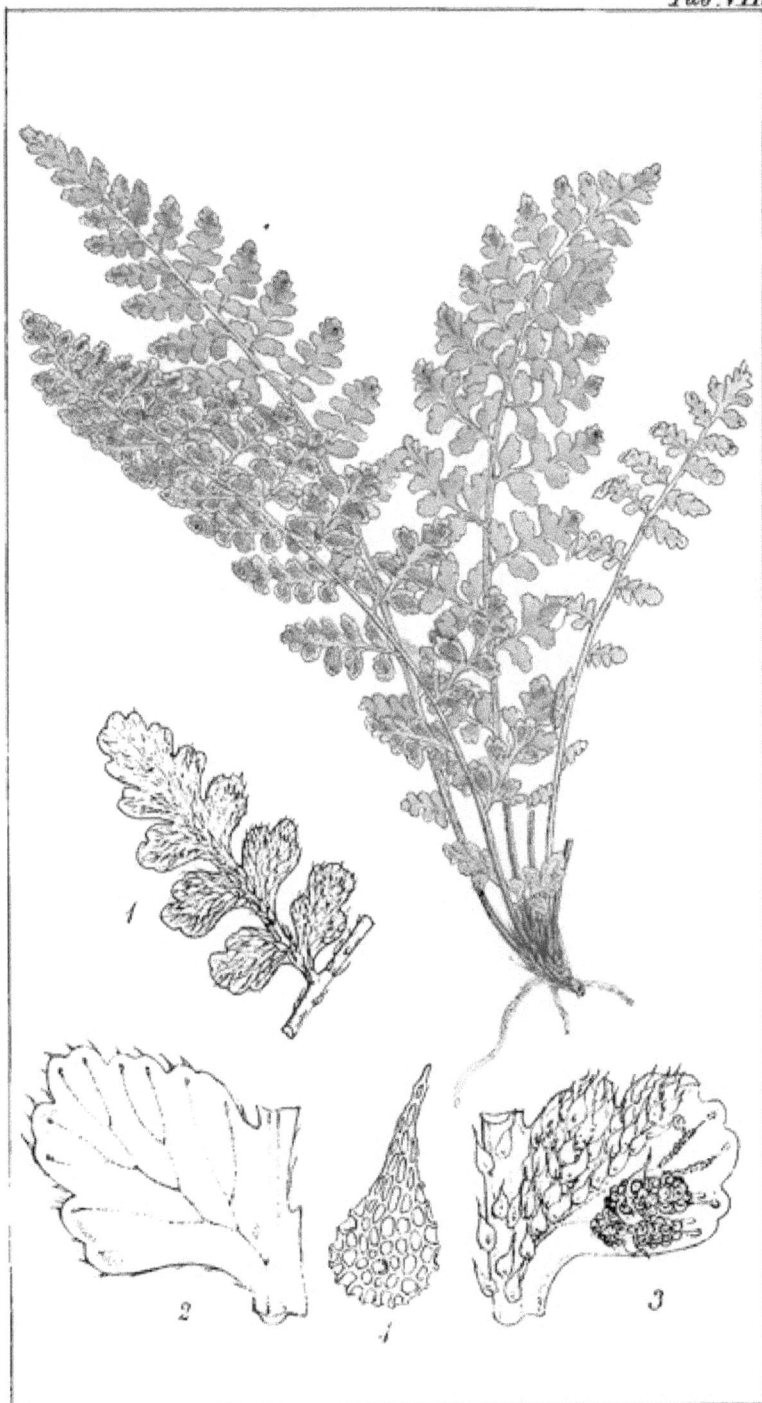

TAB. VIII.

GYMNOGRAMME PUMILA, *Spreng*

Caudice repente filiformi parce fibroso, frondibus fasciculatis 1½-biuncias longis ad basin setaceo - paleaceis sessilibus submembranaceis flabelliformi-cuneatis basi longe attenuatis superne palmatim irregulariter subdichotome incisis, segmentis acutinsculis integerrimis, venis flabellatodichotomis (costa nulla) liberis ante apicem evanescentibus, soris linearibus elongatis non raro (cum venis) dichotomis demum magis minusve confluentibus.

Gymnogramme pumila, " *Spreng. Tent. Suppl. ad Syst. Veg. p.* 31." *Kze. Analecta Pterid. p.* 11. *t.* 8. *f.* 1. *Moore, Ind. Fil. p. lxii.*

Hecistopteris pumila, *J. Sm. in Lond. Journ. of Bot.* 1 *p.* 193. *Fée, Gen. Fil. p.* 179. *t.* 16. *B.*

HAB. Tropical America, Surinam, on trees in moist woods, *Weigelt.* French Guiana, *Leprieur.* Brazil, Para, *R. Spruce, n.* 57. *and* 58. Isle of Coyba, coast of Veraguas, *Seemann.*

A very distinct and remarkable Fern; till recently, supposed to be peculiar to French and Dutch Guiana, now found in Brazil, and, still more recently even, off the Coast of Veraguas in the Pacific.

Fig. 1. Fertile frond. *f.* 2. Sterile frond. *f.* 3. Portion of a fertile frond, with sori partially removed from the receptacle :—*magnified.*

Tab. VIII.

1

2

3

TAB. IX.

GYMNOGRAMME RENIFORMIS, *Mart.*

Caudice brevi crasso basi fibroso superne copiose paleaceo, squamis subulatis ferrugineis nitidis, stipitibus cæspitosis 3-uncialibus ad semipedalem ebeneis, frondibus sesquiuncialibus coriaceis reniformi-rotundatis, venis approximatis flabellatis dichotomis, soris linearibus parallelis in zonam semilunatam discum occupantem confluentibus.

Gymnogramme reniformis, *Mart. Ic. Pl. Crypt. p.* 88. *t.* 26.
Pterozonium reniforme, *Fée Gen. Fil. p.* 178. *tab.* 16. A. *Moore, Ind. Fil. p.* lxi.

HAB. Brazil; dense woods on Mount Cupati, near the River Japura, *Martius.* Near Tarapota, Eastern Peru, on Mount Guayrapurima, *Spruce,* 1856.

One of the rarest and most beautiful of Ferns, with its undivided reniform fronds, and the curious crescent-shaped mass of sori, and the glossy ebeneous stipites. I am not aware that it has ever been gathered, but by the two eminent Botanists and S. American travellers just mentioned. Mr. Moore, in adopting the Genus *Pterozonium*, does not fail to remark, that it is "technically not very different from *Gymnogramme*, but the aspect of the plant is so peculiar, that the parallel contiguous receptacles, from which result a broad submarginal confluent sorus, may well be considered sufficiently distinctive." I am not aware that its habit is more peculiar in the genus than the *Gymnogramme* figured in our preceding plate, which, nevertheless, Mr. Moore properly retains in that genus, rather than adopt J. Smith and Fée's *Hecistopteris.* At any rate, our present Fern is not more remarkable among the species of *Gymnogramme*, than *Adiantum reniforme* and *asarifolium* and *Parishii, Trichomanes reniforme*, and *Lindsæa reniformis*, are in their respective genera.

Martius appears to have gathered specimens nearly a foot tall.

Fig. 1. One half of a fertile frond, seen from beneath. *f.* 2. Portion of do., showing the arrangement of the sori on the veins, and the receptacles :—*magnified.*

CENT. 2. T. 9.

TAB. X.

POLYPODIUM (EUPOLYPODIUM) SPRUCEI, *Hook.*

Nanum, caudice brevi subfiliformi repente, stipitibus dense
cæspitosis vix bilinearibus, frondibus membranaceis subunciam longis oblongo-subspathulatis obtusis indivisis integerrimis utrinque marginibus stipitibusque villis longis
ferrugineis basi latioribus scariosis pilosis, costa gracili flex-
- uosa, venis obliquis remotis simplicibus apice clavatis intra
marginem terminantibus, soris paucis in apicem venarum
superiorum globosis.

HAB. Near Tarapota, Eastern Peru, *Spruce, n.* 4746.

I do not find, anywhere described, a species which will agree
with this; notwithstanding, the valuable "Enumeratio Speci-
erum" of the Genus *Polypodium*, lately published by Mettenius, in which he enumerates 268 species: but, in so doing
he considers *Polypodium* in its more extended, or I may say,
Willdenovian sense; omitting, however, *Phegopteris*, of which,
he has a separate "Enumeratio," of no less importance.

Fig. 1. Back and front view of a frond; *magnified.* *f.* 2.
Apex of a fertile frond :—*more magnified.*

Tab. X.

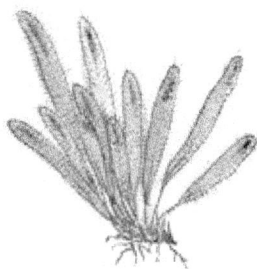
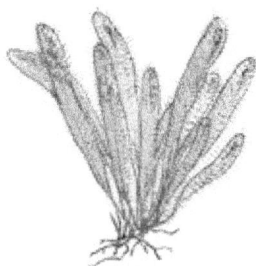

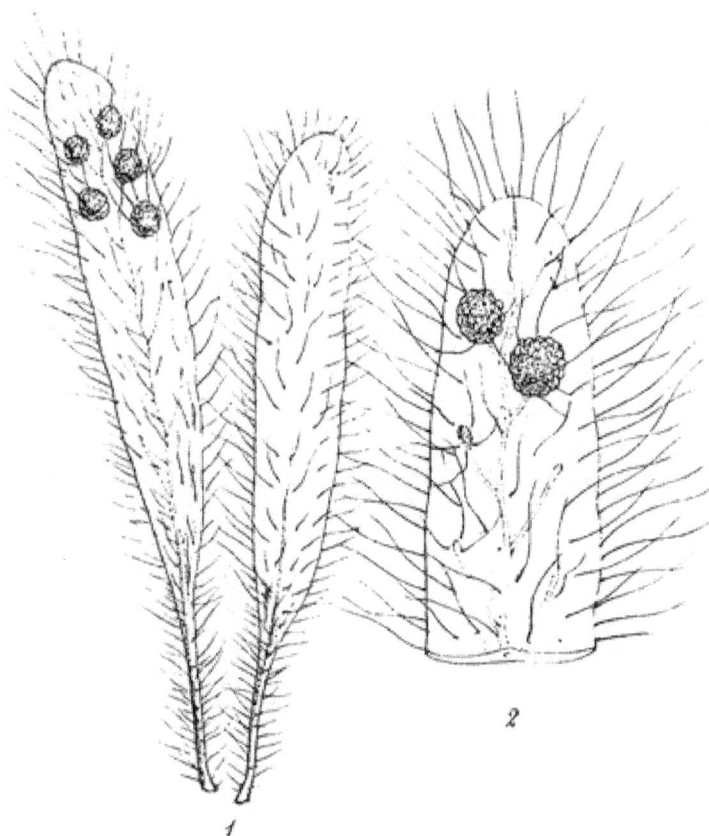

1

2

Fitch del. et lith.

Pamplin, imp

TAB. XI.

ASPLENIUM (EUASPLENIUM) TRILOBUM, *Cav.*

Parvum, caudice crassiusculo erecto ad apicem paleaceo, squamis imbricatis nitidis, stipitibus cæspitosis 2-2½ pollices longis, frondibus coriaceis rhombeis acuminatis integris crenatisve 3-lobis vel rarius 3-partitis segmentis acuminatis magis minusve incisis lobo intermedio longiore, venis immersis pluries subflabellatim dichotomis, soris majusculis erecto-patentibus lineari-oblongis, involucris rigidis fuscis.

Asplenium trilobum, *Cav. Præl.* 181, *p.* 255. *Willd. Sp. Pl.* 5. *p.* 306. *Hook. Spec. Fil. v.* 3. *ined. Gay, Fil. Chil.* 6. *p.* 499. *Metten. Asplen. p.* 146.

Asplenium trapezoides, *Sw. Syn. Fil. p.* 76. *Willd. Sp. Pl.* 5. *p.* 306. *Schk. Fil. t.* 67. *Gay, Fil. Chil.* 6. *p.* 499. *Metten. Asplen. p.* 146.

Asplenium parvulum (*small state*), *Hook. Ic., Pl. t.* 222.

HAB. S. Chili and Chiloe, *Poeppig, Cuming. n.* 820; *Captn. Ph. King, W. Lobb. Lechler, Gay, Harvey, &c. &c.* Peru.? (*Swartz*). Mariane Islands? (*Willdenow*). S. Brazil, *Tweedie, in Herb. nostr.*

A species easily recognized by its size and trapezoid form. *A. trapezoides*, indeed, is a more appropriate name than *A. trilobum*; but the latter has the right of priority. Our *Aspl. parvulum* of the Icones Plantarum, is merely a smaller and young state of this; we were the more disposed to think it distinct, from its being detected on the Atlantic side of America: having been previously known only on the Pacific coasts.

Fig. 1. Portion of a frond :—*magnified.*

Tab. XI

TAB. XII.

POLYPODIUM (EUPOLYPODIUM) TRICHOSORUM, *Hook.*

Caudice subrepente crassiusculo fulvo-crinito, stipitibus cæspitosis 2-3-uncialibus gracilibus filiformibus, frondibus indivisis 4-5-uncialibus subspathulato-lanceolatis membranaceis translucidis flavo-viridibus obtusiusculis integerrimis vel crenato-lobulatis (lobis obtusis) supra parce subtus marginibus stipiteque pilis ferrugineis patentibus longis copiose crinitis, venis patentibus flexuosis subdichotome pinnatis, venulis apice clavatis, soris sub-4-serialibus parvis in apicem venularum, pilis plurimis inter capsulas.

HAB. On Trees, forest of Archedona, Quitinian Andes, *Prof. W. Jameson, n.* 349.

This is one among the many Andinian Ferns, which, as far as my researches extend, I take to be new, and for which I am indebted to Prof. W. Jameson's zeal and kindness. It is a graceful Fern, probably pendent in its native locality, judging from the slenderness of the stipes and a peculiar curvature in many of the specimens at the setting on of the frond upon the stipes.

Fig. 1. Portion of the upper side of a frond. *f.* 2. 3 portions of the underside with sori :—*magnified.*

Tab XII

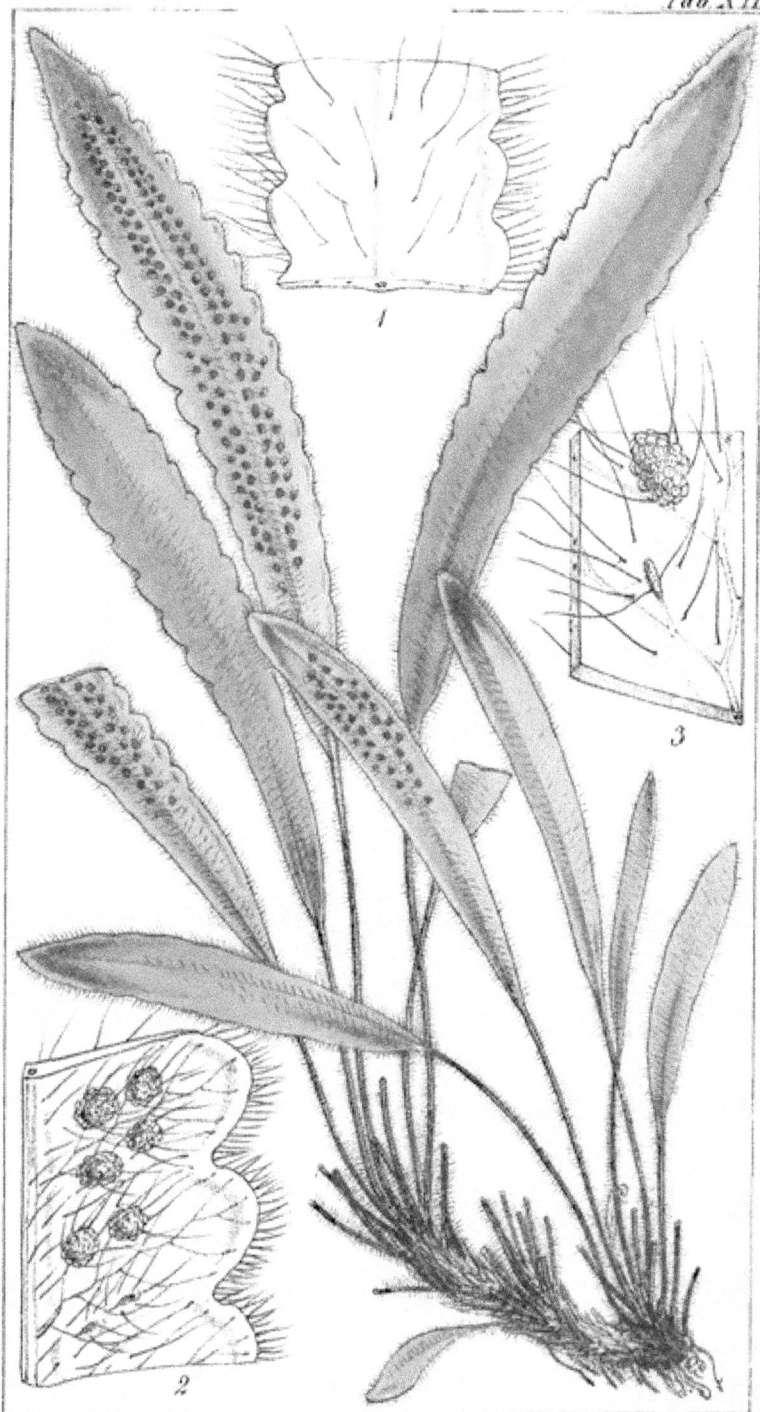

TAB. XIII.

HYMENOPHYLLUM SIMONSIANUM, *Hook.*

Caudice filiformi gracili longe repente, frondibus solitariis distantibus oblongo-lanceolatis membranaceis laxis fuscis bipinnatifidis in stipitem brevem gracilem basi attenuatis apice obtusis, lobis primariis semiuncialibus oblique cuneatis sub lente argute serratis margine inferiore truncatis integris superiore cum apice lobato-pinnatifidis, lobulis paucis (3-5) obtusissimis, involucris in lobis terminalibus frondis ovalibus subobovatisque exsertis profunde bivalvibus, valvis convexis subspinuloso-serratis, venis apice clavatis, soris inclusis recaptaculum tegentibus.

HAB. Khasya Hills, Eastern Bengal, *Simons.*

This does not appear to be anywhere described, though it must be confessed, that in so extensive a natural genus, it is very difficult, and in a few words, to define the character of any particular kind. Accurate figures are most to be depended upon.

Fig. 1. A superior fertile lobe. *f.* 2. Involucre laid open vertically shewing the sorus. *f.* 3. Entire involucre :— *magnified.*

Tab. XIII.

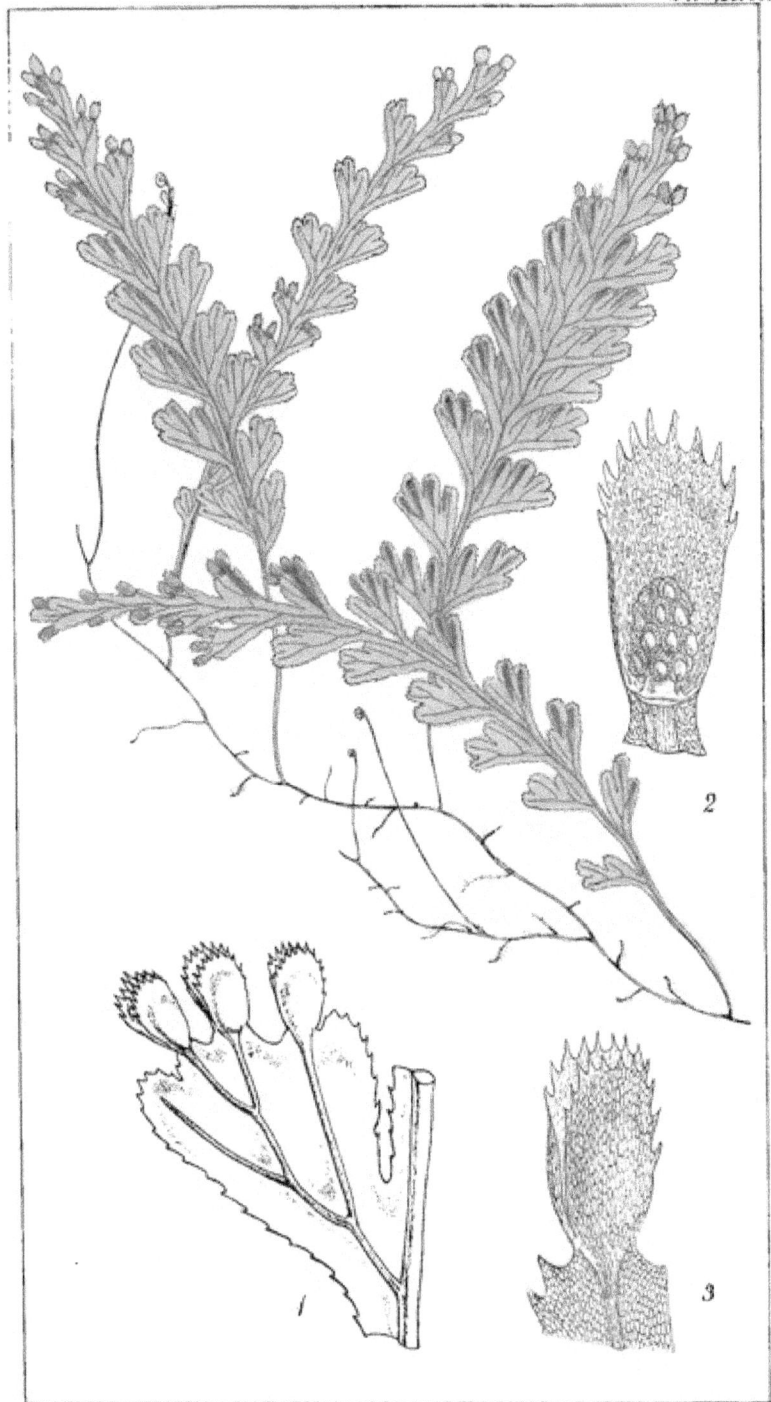

TAB XIV.

XIPHOPTERIS JAMESONI, *Hook.*

Caudice parvo (ut videtur) repente, stipitibus gracilibus subsemiunciam longis nudis, frondibus 4 pollices longis 1½
lineam latis erectis subcoriaceis pallide viridibus profunde
ad rachin fere pinnatifidis apice in caudam longam in-
tegerrimam soriferam terminantibus, lobis horizontalibus e
basi latiore oblongis·obtusis univeniis, venis internis simplicibus apice clavatis, soris in caudam terminalem linearibus
venas totas tegentibus confluentibus, capsulis longe pedi-
cellatis.

HAB. Andes of Quito, *Prof. W. Jameson.*

That the species of the Genus *Xiphopteris* are very variable, is notorious to every student of Ferns; and, I believe that the *X. myosuroides* and *X. serrulata* are now generally looked upon as forms of one species. It may, therefore, be considered a bold step to constitute a species of the
present singular and particularly neatly formed kind, which,
at first sight, and independent of its fertile cauda, has more
the appearance of some neckeroid moss, or some delicately
pinnated *Jungermannia*, than a Fern. Instead of being only
strongly "toothed" as is characteristic of *Xiphopteris* generally, it is so deeply pinnatified, nearly to the rachis, that it
might almost be called pinnate: and in an advanced state of
the plant, these segments or pinnæ, fall off partially or entirely, leaving the rachis like a long, stout, naked bristle.

Mr. Moore, whose views always deserve attention, observes,
that to him the sori of the genus appear to be produced in a
line contiguous to the mid-rib, and seem little different from
Pleurogramme. To me, they appear to be decidedly on the
lateral thickened veins, extending from the costa to the apex,
just within the margin: so difficult is it for all to see with the
same eyes, and so much is there yet to learn, in the structure
of the fructification of Ferns.

Fig. 1. Upper portion of a fertile frond. *f.* 2. Segments,
superior side. *f.* 3. Portion of a frond, inferior side. *f.* 4.
Portion of a segment of a frond with one sorus, and the
receptacle of a sorus. *f.* 5. Capsule :—*magnified.*

CENT. 2. T. 14.

Tab. XIV.

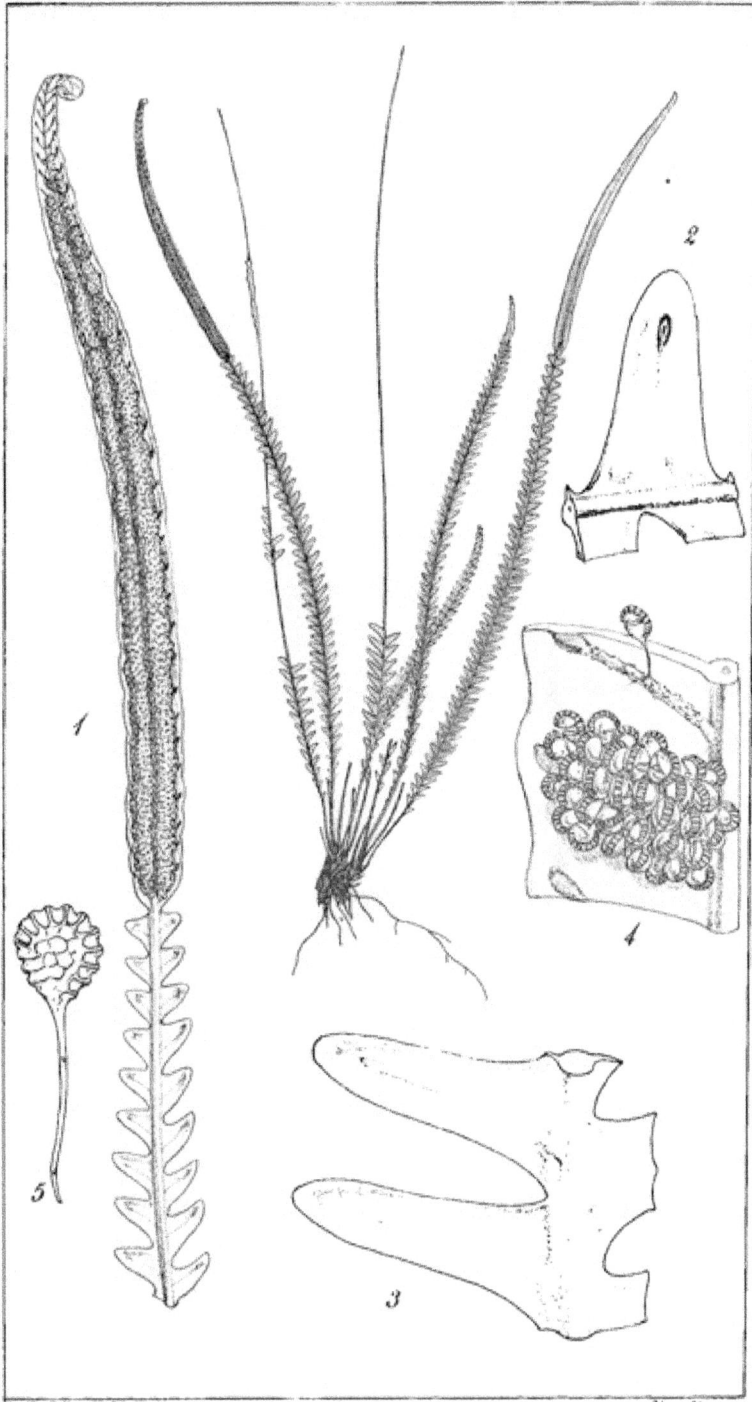

TAB. XV.

MENISCIUM PROLIFERUM, *Sw.*

Deciduo pubescens, caudice crassiusculo subrepente fibroso,
stipitibus cæspitosis erectis longitudine variantibus, frondibus pedalibus bipedalibus et ultra subcoriaceis glabris
pinnatis apicibus et in axillis pinnarum repetitim proliferis
longissime extensis, pinnis 3 - 6 uncias longis sessilibus
oblongo-lanceolatis acuminatis basi equalibus vel junioribus
præcipue inferne dilatato-rotundatis superne auriculatis
integris vel (adultis) grosse crenato-serratis, venis pinnatis,
venulis omnibus cum iis oppositis junctis et venas spurias
intermedias formantibus, soris ovalibus copiosis, singulo in
singula venula non raro confluentibus.

Meniscium proliferum, *Sw. Syn. Pl. p.* 19. *and* 207. *Willd. Sp.
Pl.* 5. *p.* 135.

Polypodium proliferum, *Roxb. Herb. Wall. Cat. n.* 312, (*not
Kaulfuss.*)

Polypodium luxurians, *Kze. in Linnæa*, 23. *p.* 280.

Phegopteris luxurians, *Metten. Phegopt. p.* 25.

Goniopteris prolifera, *Pr. Tent. p.* 183. *J. Sm. in Hook. Journ.
Bot.* 3, *p.* 396.

Ampelopteris elegans, *Kze. Bot. Zeit.* 6. *p.* 114., *Moore, Ind.
Fil. p. lxiv.; and A.* firma, *Kze. in Linnæa*, 24, *p.* 251. *Moore,
Index. l. c.*

HAB. India, *Koenig.* Nepal, Oude, Sylhet, (*Wallich,*) and
apparently all over India; from Nilghiri and N. Western
India, Khasya, and Sikhim, Himalaya in the East, *Griffith,
Hooker, fil. and Thomson, Jacquemont, n.* 1419, *Schmidt, &c.
&c.* Luzon, *Cuming, n.* 168. Java, *Zollinger.*

This is, doubtless, a very sportive plant, but its main
feature is occasioned by its extraordinary tendency to send
out new plants from axillary and terminal gemmæ, which
take root, cover a great extent of ground, and hinder the
real form and structure of the frond from being distinctly
seen. One can hardly conceive, why so many names should
have been adjudged to this plant, and even the honor of a
new genus.

Fig. 1. Base of a pinna, seen from above. *f.* 2. Base of a
fertile pinna, seen from beneath. *f.* 3. Portion of a fertile
pinna, from near the middle, seen from beneath :—*magnified.*

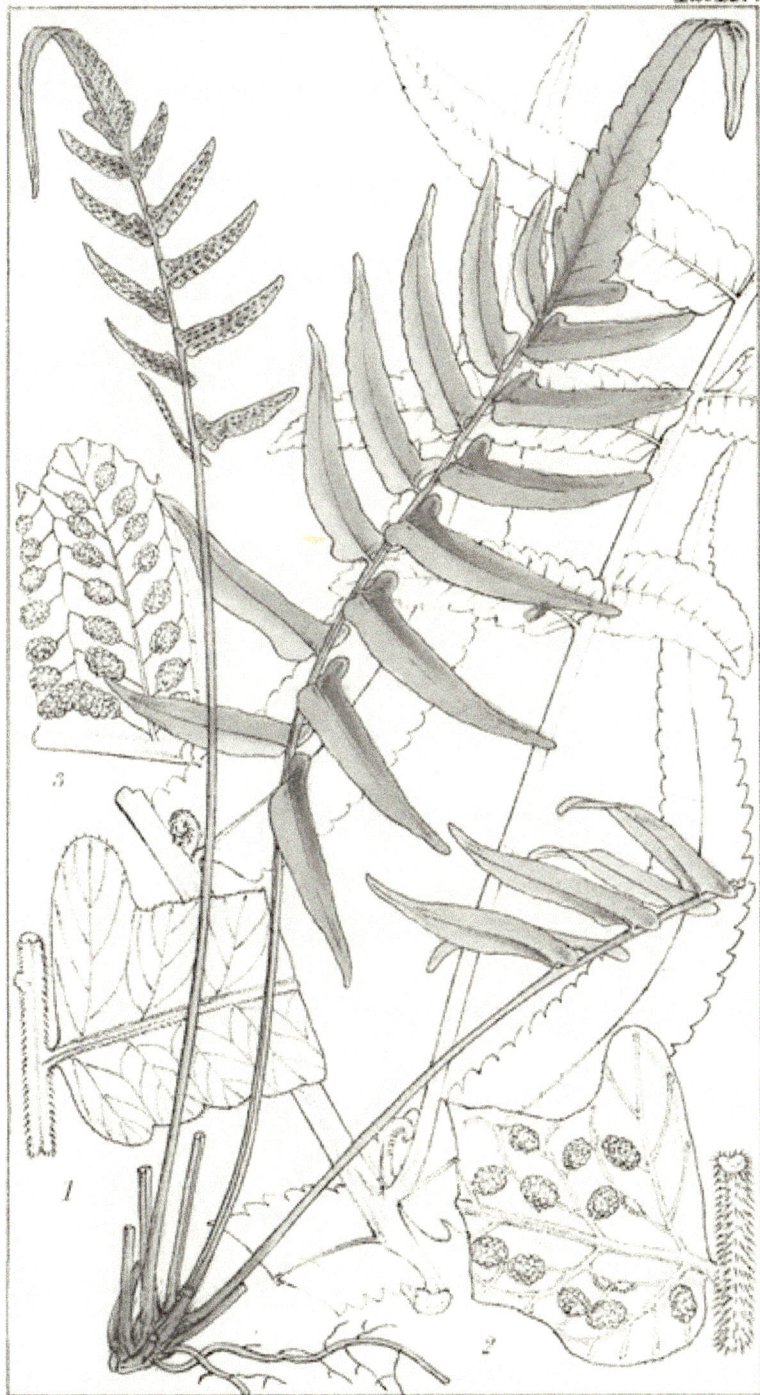

Tab XV.

Fitch del. et lith.

1

2

3

TAB. XVI.

ASPLENIUM (DIPLAZIUM) ZEYLANICUM, *Hook.*

Caudice terete repente subterraneo nudo fibroso-radicante, stipitibus sparsis solitariis squamis laxis intense fuscis lanceolato-subulatis paleaceis 4-uncias ad spithamæam longis, frondibus lanceolatis acuminatis subcoriaceo-membranaceis spithamæis ad pedalem 1-2-uncialibus latis profunde pinnatifidis basi pinnatis apice subintegris serratisve, lobis pinnisque patentibus oblongis obtusissimis integerrimis, venis remotis furcatis, soris linearibus, involucris venula basi superiore loborum præcipue diplazioideis.

HAB. Ceylon, banks of a large stream of Kotmalee Oija, elev. 4,000 feet, *Gardner, n.* 1249. I have received a specimen also, from *Mr. Thwaites*, but possibly from Mr. Gardner's collection.

I possess only two specimens of this distinct looking diplazioid *Asplenium*, and venture to constitute a species of it though it may possibly be a young form (yet in very good fructification) of some compound species, which time, and further researches only can show.

Fig. 1. Sterile lobe. *f.* 2. Fertile lobe:—*magnified.*

Tab.XVI

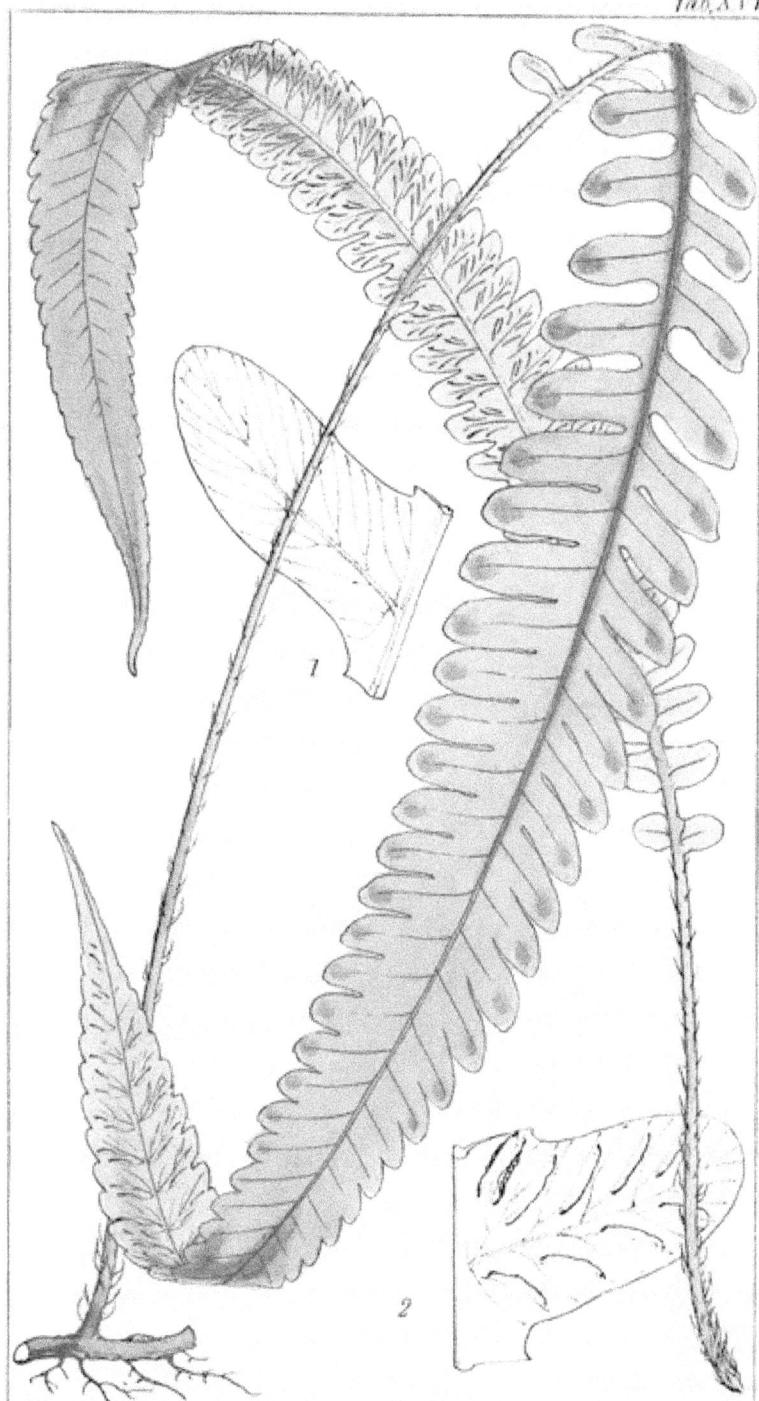

1

2

TAB. XVII.

ASPLENIUM (DIPLAZIUM) LOBBIANUM, *Hook.*

Fronde sesquipedali ovato-acuminata subcoriaceo-membra-
nacea pinnata, pinnis 5-6-uncias longis 1½ unciam latis
petiolatis patentibus oblongis acuminatis paululum falcatis
basi superiore truncata rachi parallela rarius subauriculatis,
inferiore cuneata vel subrotundata marginibus serratis, pinnis
supremis sensim minoribus confluentibus, venis subhorizon-
talibus fasciculatis 1-terve dichotomis, soris solitariis (asplen-
ioideis) vel geminatis (diplazioideis) non raro subscolopen-
drioideis nec costam nec marginem attingentibus.

HAB. Java, *Thos. Lobb.* 1846.

This may possibly be included among the 12 species of
pinnated *Diplazium* of Java, enumerated by Blume : but the
characters of that author are so very brief, and unaccom-
panied by figures, which are such great helps in critical
species, that I had better avoid all allusion to them, and con-
tent myself with saying, that among my own numerous
diplazioid *Asplenia*, there is none to which I can refer this :—
nor can I find that it corresponds with any one known to me
of other countries.

I possess but one frond, and that without caudex or stipes.

Fig. 1. Lowest pinna, *nat. size.* *f.* 2. Portion of a fertile
pinna, *magnified.*

Tab **XVII**

TAB. XVIII.

ASPLENIUM (DIPLAZIUM) CRASSIDENS, *Fée*.

Stipite stamineo-fusco spithamæo nudo, fronde subpedali coriaceo-membranacea atro-viridi subtus pallidiore opaca pinnata, pinnis paucis remotis 10-11 brevi-petiolatis patentibus remotis 4-uncialibus inæquilaterali-ovato-lanceolatis basi oblique attenuato-cuneatis subito acuminatis subgrosse serratis præcipue supra medium, supremis subdecurrenti-sessilibus ultimis tribus in unica hastiformi confluentibus, venis approximatis furcatis v. bis-terve dichotomis magis ramosis in pinna terminali, soris obliquis parallelis æqualibus e costa ad marginem attingentibus non raro præcipue ad apicem frondis diplazioideis, involucris fuscis.

Diplazium crassidens, *Fée*, 8me. *Mém. p.* 82. *Metten. Asplen. p.* 151.

HAB. N. Grenada, Paramos of Ocaña, elev. 10,000 feet, *Schlim, n.* 393.

A peculiar looking species, with a costa dividing the pinna into 2 unequal portions, the upper half being the broadest. It is remarkable for the great regularity and length of the brown sori, extending, as they do, from the costa to the margin.

Fig. 1. Terminal fertile pinna seen from beneath :—*nat. size.*
f. 2. Portion of the same with sori :—*magnified.*

Tab.XVIII

Fitch del et lith.

Panoplin, imp.

TAB. XIX.

ASPLENIUM (DIPLAZIUM) FRAXINIFOLIUM, *Wall.*

Caudice crasso declinato vix repente copiose crasse fibroso apice nigro-paleaceo, stipitibus aggregatis sæpe pedalibus fuscis subrobustis parte inferiore præcipue laxe et deciduo nigrescente-paleacis, frondibus pedalibus ad sesquipedalem subcoriaceis firmis subnitentibus siccitate fuscis pinnatis, pinnis remotis 3-11 petiolatis patentibus 6-8-10-uncialibus late oblongo-lanceolatis tenui-acuminatis integerrimis basi suboblique cuneatis, venis fasciculatis bis terve dichotomis parallelis copiosis omnibus liberis, soris copiosis lineari-elongatis a costa fere ad marginem continuis, involucris angustis.

Asplenium fraxinifolium, *Wall. Cat. n.* 194.
Diplazium fraxinifolium, *Wall. Herb.* 1823. *Moore, Index Fil. p.* 133.
Diplazium elegans, *Hook. in Florul. Hong-Kong. Kew Gard. Misc., var. venis liberis. C. Wright, in Herb. of U. S. Expl. Exped., under Commodores Ringgold and Rodgers, and in Herb. Nostr.*

HAB. Penang, *Wallich.* Sincapore, *Thomas Lobb. n.* 33. Khasya, *Griffiths.* Assam, *Griffiths, Simons, Hooker fil. et Thomson.* Hong-Kong, *Mr. Alexander, Dr. Harland, Wilford, C. Wright.*

This fine Fern has been largely distributed by the late Dr. Wallich, under the name of *Asplenium* (Diplazium) *fraxinifolium;* and I am desirous of making it known to the botanical world ; for as far as I am aware, it has been noticed by no Fern-author in any way, save that, in Moore's Index Filicum, Wallich's *Asplenium fraxinifolium* is called "*Diplazium fraxinifolium.*" My impression is that this Fern, with its free venation will prove specifically identical with *Diplazium* (Oxygonium) *elegans*, figured in Hook. Ic. Pl. t. 939-940, notwithstanding the anastomosing venation of the latter. I would observe, too, that the *Diplazium alternifolium*, Bl. (and Hook. Fil. Exot. Tab. 17) has, though possessing much broader pinnæ, great affinity with this ; but the whole group of diplazioid *Asplenia* require a careful study and revision, which we trust to attempt ere long.

Fig. 1. Portion of a fertile frond seen from beneath:— *magnified.*

Tab. XLX.

TAB. XX.

Asplenium (Euasplenium) quitense, *Hook.*

Parvum, caudice filiformi repente hic illic inferne radiculoso,
stipitibus subfasciculatis 3-4 ex eodem puncto 1½ unciam
longis nudis gracilibus herbaceis superne una cum rachibus
marginato-alatis, frondibus 2½-3½ uncias longis submembranaceis atroviridibus lanceolatis pinnatis, pinnis 6—9-jugis
horizontaliter patentibus sublonge petiolatis oblique ovatis
pinnatifido-lobatis basi inferne cuneato-incisis, lobis obtusis
integerrimis basi superiore lobo majore auriculæformi bi-
trifido, venis simplicibus vel (in auriculam) bi-trifurcatis
apice intra marginem clavatis, soris majusculis oblongis,
involucro membranaceo latiusculo.

Asplenium Quitense, *Hook. Spec. Fil. vol.* 3. *ined.*

Hab. On decayed trees in the forest of Archedona, Quitinian Andes. *Prof. W. Jameson, n.* 707.

I refer this distinct and very pretty Fern, to the *Trichomanes*—division of *Euasplenium*, and it has some affinity with
several of the less rigid, herbaceous species of that section ;
but is peculiar in the slightly winged rachis, the very distinctly petiolated, rather deeply pinnatifid, lobate pinnæ, and,
above all, in the long filiform creeping caudex.

Fig. 1. Sterile pinna. *f.* 2. Fertile pinna, seen from
beneath. *f.* 3. Portion of a pinna, with sorus :—all *magnified.*

Tab. XX.

TAB. XXI.

Polypodium (Eupolypodium) tenuisectum, *Bl.*

Caudice "repente sublignoso," stipitibus solitariis erectis flexuosis pilis setisve ferrugineis patentibus hispidissimis, frondibus digitalibus ad dodrantalem lato-lanceolatis acuminatis rigido - subcoriaceis sparse setaceo-pilosis, pinnis primariis remotiusculis 2-pollicaribus lineari-acuminatis patentibus, pinnulis 1¼ lineam longis sessilibus linearibus acutiusculis integerrimis inferne subdecurrentibus univeniis, venis dimidiæ pinnarum longitudine apice clavatis, soris solitariis globosis ad basin venarum insertis, rachi compressa.

Polypodium tenuisectum, *Bl. En. Fil. Jav. p.* 134. *Fil. Jav. p.* 189. *tab.* 88. *A. Mett. Fil. Lechl. p.* 5. *tab.* 2 *f.* 1—3. *Metten. Polypod. p.* 54.

Polypodium myriophyllum, *Mett. l. c. p.* 6.

Hab. In clefts of rocks and on old trees, on the lofty mountains of Java, *Blume.* Trunks of trees, near Tatanara, Peru, *Lechler.*

The only localities known of this elegant and rare species of *Polypodium*, are the high mountains of Java, and those of the Peruvian Andes. I possess, indeed, only Peruvian specimens; but they quite accord with Blume's figure.

Fig. 1. Portion of a fertile pinna with pinnules. *f*, 2. Sterile pinna:—*magnified.*

Tab. XXI

TAB. XXII.

ASPLENIUM (EUASPLENIUM) GIBERTIANUM, *Hook.*

Caudice crasso descendente copiose radiculoso superne setoso-squamoso, stipitibus numerosis cæspitosis vix unciam longis compressis viridibus demum dorso præcipue castaneis superne alatis, frondibus subsemipedalibus lanceolatis membranaceis pulchre viridibus pinnatis, pinnis numerosis approximatis $\frac{1}{2}$-$\frac{3}{4}$-unciam longis patentibus ovato-lanceolatis profunde pinnatifidis sessilibus basi oblique cuneatis et in alam latiusculam longe (usque ad insertionem) pinnæ adjacentis inferioris profunde pinnatifidis, lobis oblongo-lanceolatis acutissimis integerrimis infernis subcuneatis biquadrifidis pinnis infimis subflabelliformibus, venis simplicibus vel in laciniis 2—4-fidis bi-furcatis, soris oblongis discoidalibus (e margine et costa remotinsculis), involucro membranaceo albido, rachi insigniter compressa herbacea apice sæpius longe excurrente prolifera.

Asplenium Gibertianum, *Hook. Spec. Fil. vol. 3. ined.*
Asplenium inciso-alatum, *Moore, MS. in Herb. Hook. and in Index Fil. p.* 137. (*name only*).

HAB. Assumption, State of Paraguay, (not Island of Assumption, as given by Mr. Moore), *M. Gibert.*

A very delicate and beautiful species, of which I have seen only one fine specimen, kindly sent me by M. Gibert, a gentleman chiefly resident at Monte Video, but who has contributed much to our knowledge of the Nat. History of Paraguay, and to whom I desire to dedicate the species. It belonging to a family of plants, of which there are comparatively few representatives in Paraguay, judging from the proportion of them with other plants that have yet come to us. It will rank near the well-known *Aspl. cicutarium.*

Fig. 1. Sterile pinna and winged rachis. *f.* 2. Fertile pinna and winged rachis. *f.* 3. Bifid segments of a pinna and sorus:—*magnified.*

Tab. XXII

TAB. XXIII.

DAVALLIA (CUNEATÆ) GOUDOTIANA, *Kze.*

Pumila, caudice repente gracili, stipitibus remotis solitariis unciam sesquiunciam longis gracilibus basi parce squamosis, frondibus 3-4-uncialibus lanceolatis paululum acuminatis membranaceis pinnatis glaberrimis, pinnis subsessilibus profunde subtripinnatifidis ⅜ unciam longis, laciniis omnibus cuneatis integris vel bifidis obtusis, venis seu costis simplicibus v furcatis, soris solitariis in apicibus segmentorum utrinque vel uno latere dente instructis involucrum paullo excedente, involucris reniformibus bivalvibus.

Davallia Goudotiana, *Kze. in Annal. Pterid. p.* 35. *t.* 22. *f.* 2. *Hook. Sp. Fil.* 1. *p.* 188. *tab. L. C.*

HAB. Madagascar, *Goudot, Dr. Lyall,* and *Bojer, in Herb. Nostr.*

Having received from the late Professor Bojer of Mauritius, more perfect specimens than I was possessed of, when publishing the first vol. of my "Species Filicum," I gladly publish figures of them on the present occasion.

Fig. 1. Portion of a fertile frond. *f.* 2. Segments with sori and involucres :—*magnified.*

Tab. XXIII.

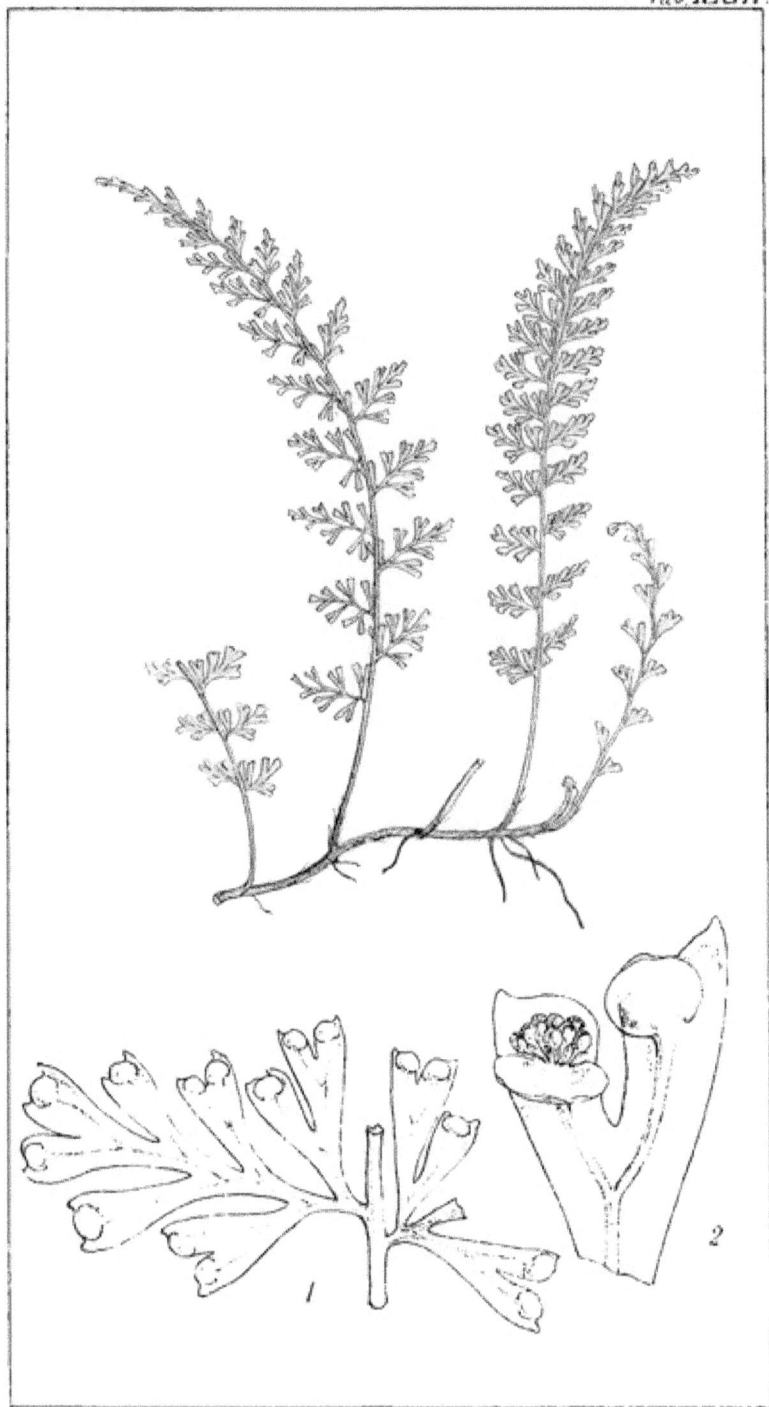

1

2

TAB. XXIV.

POLYPODIUM (PHEGOPTERIS) DAREÆFORME, *Hook.*

Caudice crassiusculo brevi subrepente dense ferrugineo-pa-
leaceo squamis lanceolatis acuminatissimis, stipite 4-unciali
nitido pallide castaneo, fronde spithamæa ovato-deltoidea
submembranacea bipinnata, pinnis primariis 4-5 uncias longis
1¼ unciam latis oblongo-lanceolatis subsessilibus acuminatis,
pinnis oblongo-ovatis subbipinnatifidis laciniis obovato-
linearibus obtusissimis simplicibus vel bifidis, venis seu costis
in divisionibus solitariis ante apicem terminantibus clavatis,
soris solitariis parvis ad basin laciniarum ultimarum, capsulis
in utroque soro perpaucis.

HAB. Khasya hill, *Simons, n.* 98.

This finely cut *Polypodium* has some affinity with *Polypod.
tenuisectum*, Bl. and perhaps, still more, with Blume's *P.
millefolium*, but besides other characters, the form of the two
is quite different, and the ramification is much more compound.

Fig. 1. Secondary pinna with sori. *f.* 2. Portion of a
pinna. *f.* 3. an ultimate segment with a sorus, from which
most of the few capsules are removed :—*magnified.*

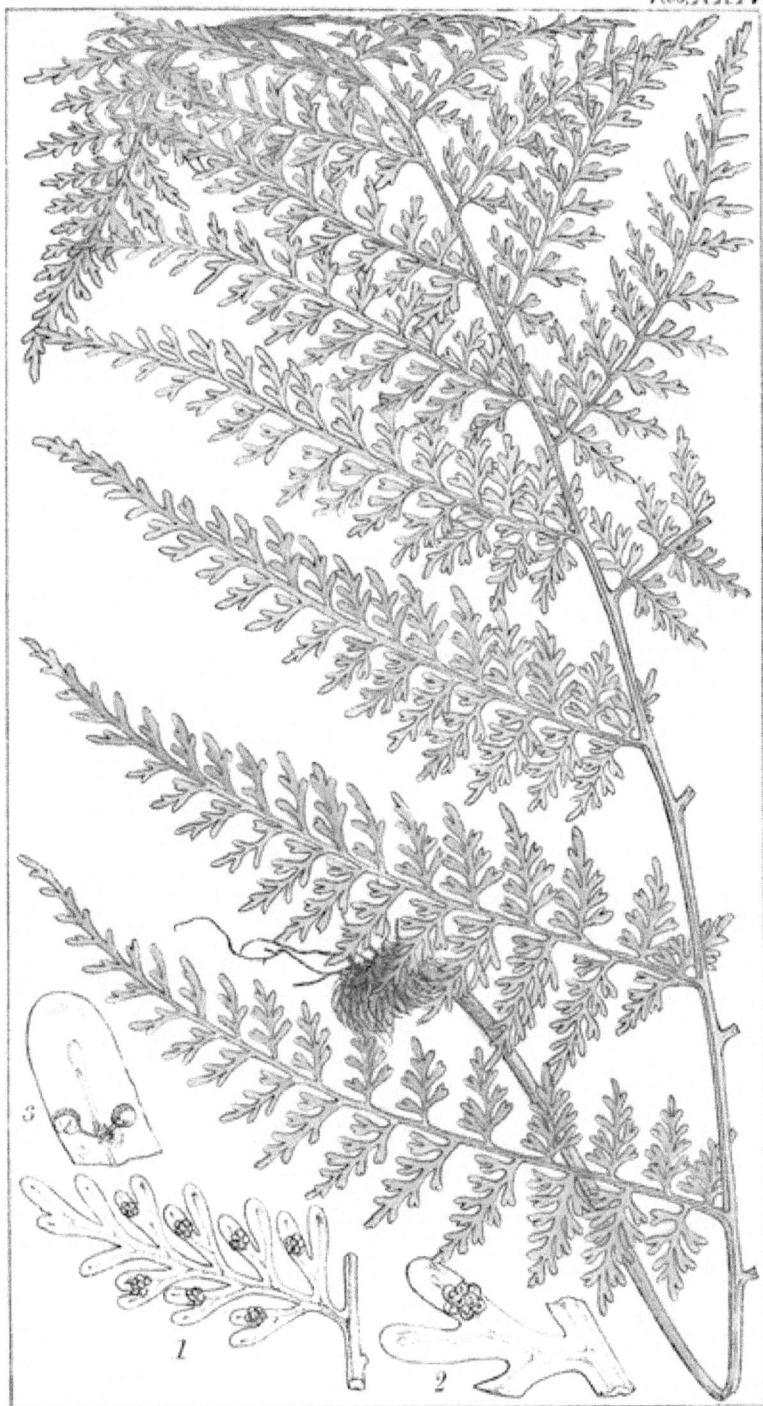

Tab. XXIV.

TAB. XXV.

Aspidium (Polystichum) Thomsoni, *Hook.*

Caudice brevi crasso obliquo apice squamoso, stipitibus basi
insigniter squamosis 3-4-uncialibus gracilibus stramineis
una cum rachi straminea setoso-paleaceis, frondibus digitalibus ad spithamæam lanceolatis acuminatis subchartaceo-
membranaceis sessilibus pinnatis, pinnis 3-4 lineas longis
sessilibus ovatis profunde pinnatifidis præcipue ad marginem
superiorem vel bipinnatis, pinnulis ovatis lobisque spinu-
loso-incisis serratisve, venis pinnatis venulis ultimis clavatis,
soris solitariis in singula pinnula seu lobo majusculis dorso
venulæ insertis, involucro subovato membranaceo peltato
pedicellato margine sæpe eroso.

HAB. Sikkim Himalaya, *Hooker, fil. et Thomson.* Above
Simla, *Col. Bates.* Kamaon, elev. 12,000 feet, *Strachey
and Winterbottom, n.* 9.

This must rank among the smallest of the *Polystichum*-
group of *Aspidium*, approaching nearest to the smallest specimens of Dr. Wallich's *Aspid.* (Polystichum) *oxyphyllum.* The
involucre, if constant to its form and pedicel, is very remarkable.

Fig. 1. Fertile pinna, almost again pinnate. *f.* 2. Portion
with sori. *f.* 3-4. Involucres : *magnified.*

Tab. XXV

TAB. XXVI.

ACROSTICHUM (GYMNOPTERIS) LINNÆANUM, *Hook.*

Caudice repente squamoso subtus fibroso, stipitibus sparsis approximatis 2-6-uncialibus inferne squamis nigris subulatis
parce paleaceis, frondibus subdimorphis; *sterilibus* 4-8 uncias
longis semiunciam latis submembranaceis elongato-lanceolatis subopacis superne sæpe longe acuminatis apice radicantibus et proliferis margine integerrimis, venis indistincte
pinnatis, venulis primariis transversis secundariisque varie
anastomosantibus, areolis majusculis subhexagonis rarissime
appendiculatis, stipitibus 2-3 uncias longis; *fertilibus* duplo
angustioribus rigidioribus 4-uncialibus lineari-lanceolatis.

Leptochilus Linnæanus, *Fée, Acrostich. p.* 87. *tab.* 47. *f.* 2.
  *excluding probably all the synonyms; certainly all references
  to figures.*
Dendroglossa Linnæana, *Fée, Gen. Fil. p.* 81.

HAB. Malay Islands, Java, "*Zollinger, n.* 1441." Borneo,
  Motley, n. 427.

There can, I think, be no doubt of this pretty Fern being
the *Leptochilus Linnæus* of Fée, l. c.; but he quotes Linnæus'
Acrostichum lanceolatum, Amœn. Acad. 1. p. 268; though
it is quite clear that Linnæus had quite another Fern in
view, since he says of it, in Sp. Plant p. 1523, " fructificationes
sunt puncta confertissima, versus apicem frondis," and he refers
to Hort. Malab. 12. t. 27. But Swartz, long ago, showed
that the Linnæan plant was the *Polypodium acrostichoides* of
Forst. Prodr. now generally referred to *Niphobolus.*

Fée afterwards in his Genera Filicum refers Linnæus' *Acrostichum lanceolatum* to *Dendroglossa;* while Moore places it
in his *Gymnopteris,* among the *Pleurogrammeæ.*

Fig. 1. Portion of a sterile frond to show the venation,
(where however, the primary *pinnated* veins are scarcely sufficiently distinctly represented). *f.* 2. Portion of a fertile
frond seen from beneath, one side with the capsules removed.
f. 3., Capsule: *magnified.*

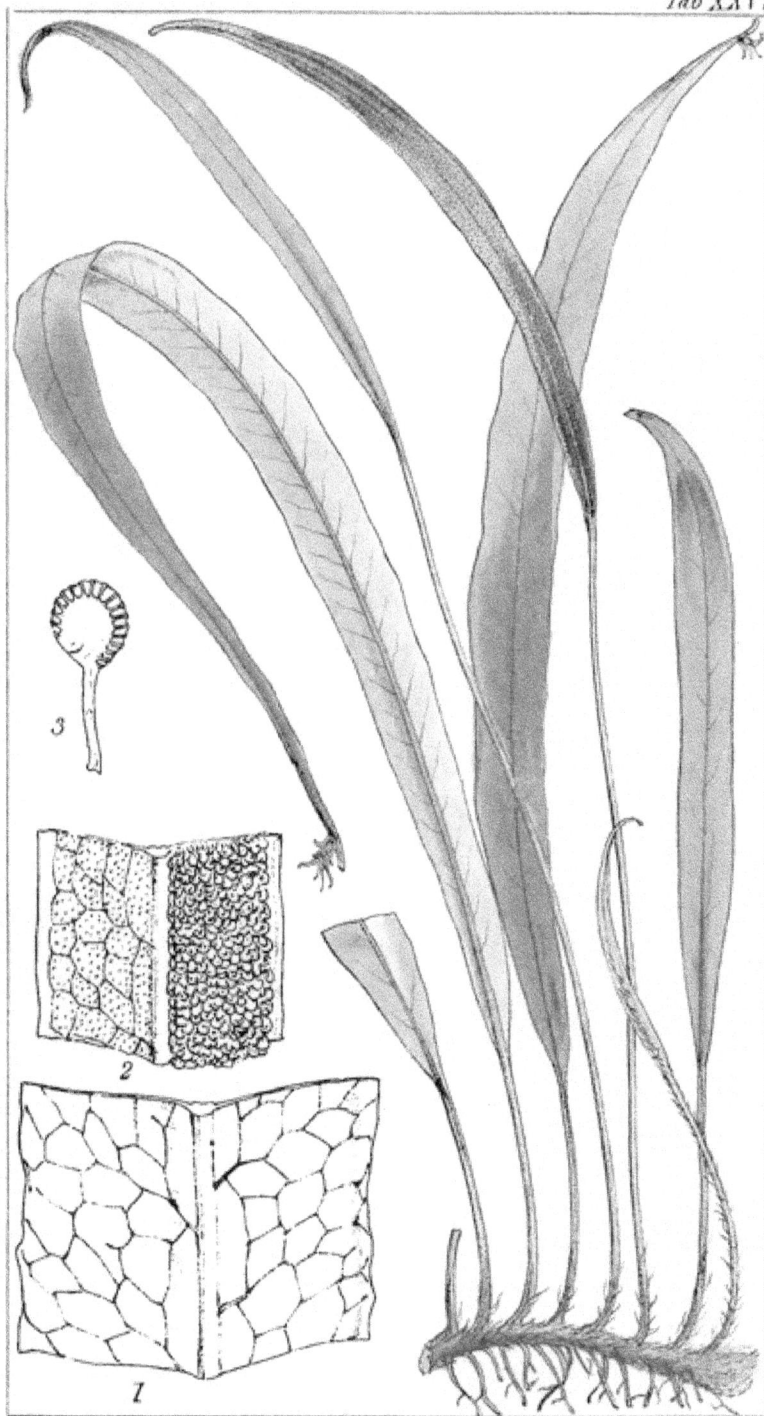

Tab XXVI

TAB. XXVII.

Asplenium (Euasplenium) Fadyeni, *Hook.*

Caudice longe repente paleaceo-squamoso radicante, fibris
longis flexuosis, stipitibus numerosis sparsis gracilibus
2-4 uncias longis inferne squamis ovatis fuscis paleaceis,
frondibus 4-6 uncias longis late ovato-lanceolatis membranaceis flaccidis bipinnatis siccitate atro-viridibus pinnis
(16-17) 1-1½ unciam longis horizontaliter patentibus remotiusculis lanceolatis basi pinnatis apicem versus pinnatifidis, pinnulis 3-5 parvis 2-3 lineas longis petiolulatis
obovato-subrhomboideis obtusis nunc profunde trilobis
lobis obovatis dentatis, venis furcatis, soris paucis oblongis parvis, involucris brunneis laxe membranaceis subathyroideis, rachi gracili subflexuosa.

Asplenium Fadyeni, *Hook. Spec. Fil.* 3. *p.* 193.

Hab. Jamaica, *MacFadyen.*

Some years ago I had the pleasure to receive this new
Fern from my late friend Dr. MacFadyen, and only upon that
one occasion. It may therefore be considered a rare *Asplenium*,
and assuredly a very distinct one, not likely to be confounded
with any other.

Fertile plant:—*nat. size. Fig.* 1. Superior basal pinna:
magnified. f. 2. Sorus: *more magnified.*

Tab. XXVII

TAB. XXVIII.

ASPLENIUM (EUASPLENIUM) ELEGANTULUM, *Hook.*

Caudice (ætate) subrobusto horizontali vel declinato ad apicem
paleaceo squamulis subulatis fuscis, stipitibus aggregatis
brevibus 1-2-3-uncialibus rachique submarginata viridibus,
frondibus 4-pollicaribus ad spithamæam elongato-lanceolatis acuminatis membranaceis flavo-viridibus bipinnatis
inferne augustatis cum pinnis nanis flabelli-vel reniformibus, reliquis ½ ad unciam longis ovatis seu ovato-
lanceolatis subsessilibus iterum pinnatis basin versus præcipue apice pinnatifidis, pinnulis ovatis subrhomboideis
obovatisve 1-1½ lineam longis acute denticulato-serratis,
venis furcatis subpatentibus in pinnis infimis subflabellatis,
soris copiosis demum confluentibus, involucris pallidis oblongis laxis subathyroideis.

Asplenium elagantulum, *Hook, Sp. Fil.* 3. *p.* 190.

Aspl. lanceolatum? *var.* elegans, *Hook Florula Hong-Kong.
in Kew Gard. Miscell.* 9. *p.* 342. *Metten. Asplen. p.* 141.

Athyrium fontanum, *Eaton, in Asa Gray's Bot of Japan,*
vi. *N. Ser. of Mem. Acad. of Arts & Sc. p.* 421 & 436.

HAB. Island near Chusan, *Alexander ;* Port Hamilton, and
Tsus Sima, Strait of Korea, *Wilford, n.* 753 ; Japan ;
Nangasaki, *Miss Nelson, Babington ;* Hakodadi, *Dr. Baines ;*
Simoda, *C. Wright.*

Late events have contributed to the opening up to us of
the Botany of N. China and Japan, whose geographical
position naturally led us to expect European forms of Ferns.
The present species was at first considered by me to be a
state of *Aspl. lanceolatum,* while Mr. Eaton looked upon it
as more related to *Aspl. fontanum.* More copious specimens
have satisfied me it is truly distinct from both. The young
fronds are simply and closely pinnated, with obliquely ovate
pinnæ.

Fertile Fronds :—*nat size. Fig.* 1. Pinnule, and *f.* 2.
Sorus : *magnified.*

CENT. 2. T. 28.

Tab. XXVIII

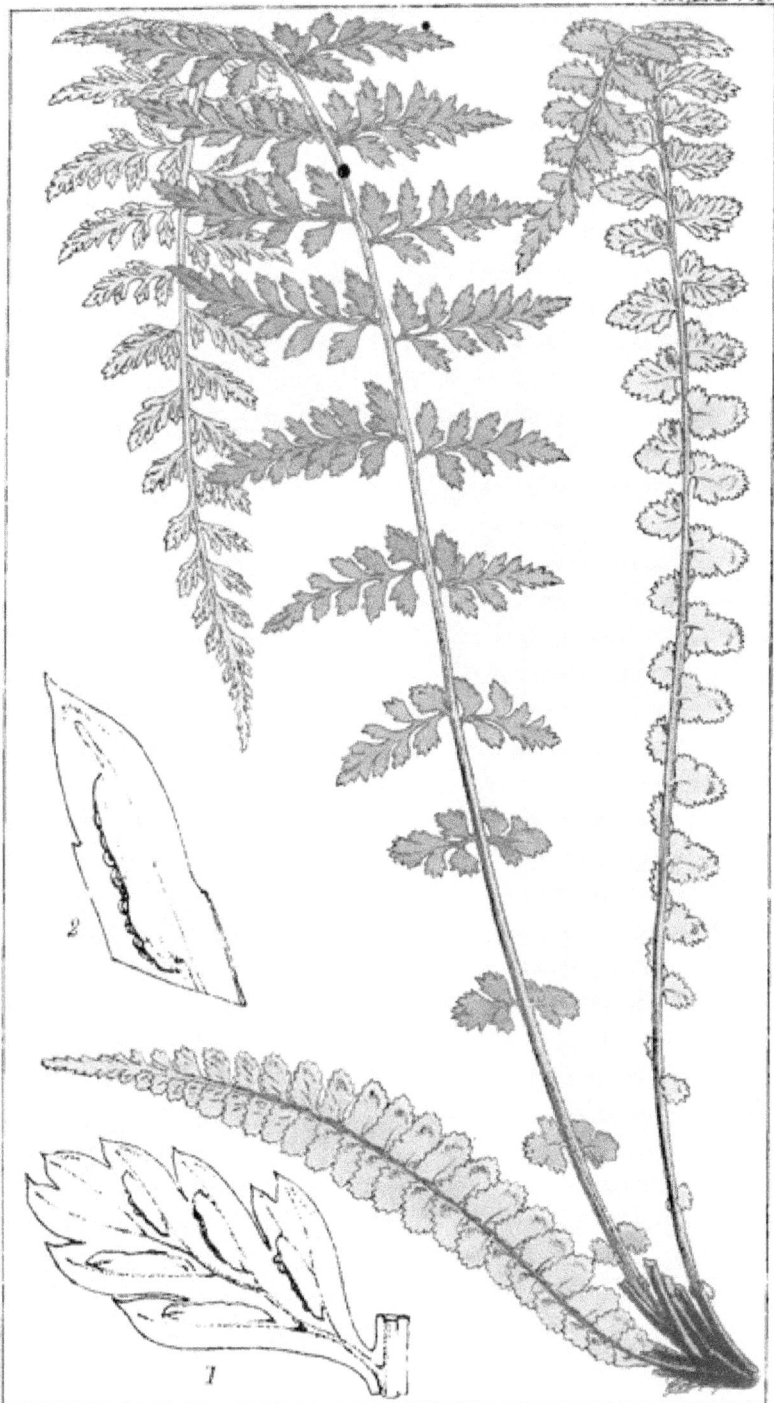

TAB. XXIX.

ASPLENIUM (EUASPLENIUM) TENUIFOLIUM, *Don.*

Caudice horizontali crassiusculo ætate vix paleaceo, stipitibus
cæspitosis 3-4 uncias ad spithamæam longis ad basin
castaneis, frondibus oblongo-ovatis acuminatis pallide viridibus membranaceis 6-12 uncias longis 3-pinnatis, pinnis
pinnulisque petiolatis, pinnis primariis 2-3 uncias longis
patentibus lato-lanceolatis acuminatis, pinnulis ultimis
obovato-v-lineari-cuneatis bi-trifidis laciniatisve, segmentis
acutissimis subspinulosis, frondis pinnarumque apicibus
pinnatifidis segmentis linearibus, venis in segmentis solitariis longe ante apicem terminantibus, soris solitariis vel
binis in quoque segmento.

Asplenium tenuifolium, *Don Prodr. Fl. Nep. p.* 8. *Kze. in
Linnæa.* 24. *p.* 265. *Metten. Asplen. p.* 128. *Hook. Sp. Fil.*
3. *p.* 194.

Asplenium concinnum, *Wall. Cat. n.* 216.

HAB. India, Nepaul, *Wallich*; Neilgherries, *Sir F. Adam, Wight,*
n. 104, *Gardner, Schmid*; Sikkim Himalaya, *Hooker. fil.
and Thomson ;* Myrung, and Mishmee, and Khasya, *Griffith ;* Ceylon, *Gardner,* n. 1079, *Thwaites* (elev. 7000 feet)
n. 3628.

An elegant, very compound species, sometimes almost
quadripinnate, allied to the West Indian *Aspl. cicutarium,* yet
very distinct, especially in the very acute segments of the
pinnules, and in the involucres never opening at or near the
margin.

Fertile Plant:—*nat. size. Fig.* 1. pinnule. *f.* 2. Sorus:
magnified.

Tab. XXIX

TAB. XXX.

ASPLENIUM (EUASPLENIUM) HALLII, *Hook.*

Caudice crassiusculo ascendente, stipitibus cæspitosis ebeneis
nitidis 1-2 uncias longis, frondibus 6-12 uncias longis submembranaceis fusco-viridibus oblongis lanceolativse basi
attenuatis apice longe acuminatis flagelliformibus suba-
phyllis radicantibus subbipinnatis seu pinnato-pinnatifidis,
pinnis primariis sessilibus ovato-lanceolatis obtusis ¾ ad
unciam longis horizontalibus subpectinato-pinnatifidis, segmentis linearibus obtusis ad basin superiorem subauriculatis
bifidis vel inferne iterum pinnatis, pinnis infimis frondis
nanis, venis pinnatis solitariis in quoque segmento, soris
parvis oblongis costam versus, involucris membranaceis,
rachi ebenea.

Asplenium Hallii, *Hook, Sp. Fil.* 3. *p.* 202.

Asplenium pectinatum, *Moore, mst. in Herb. Hook. et in Ind.
Fil. (name only) not of Wallich, nor of Mettenius.*

HAB. Forest of Esmeraldas, Ecuador, *Col. Hall;* Sao Gabriel,
valley of the Amazon, on young trees and shrubs, "fronds
spreading horizontally," (no doubt rooting at the extremity),
Spruce, n. 2357.

No other stations than the above have yet been recorded for
this rare species. In the flagelliform and radicant apex it
resembles *Aspl. rhizophyllum,* but the pinnæ and pinnules
are widely different. It is remarkable for the ebeneous short
stipes and rachis, and the dwarfed lower pinnæ extending
almost to the caudex.

Fertile plant: *nat. size. Fig.* 1. portion of a bipinnate
form: *nat. size. f.* 2. primary pinna, fertile, and *f.* 3. involucre:
magnified.

Asp. 2 1. 30

Tab XXX

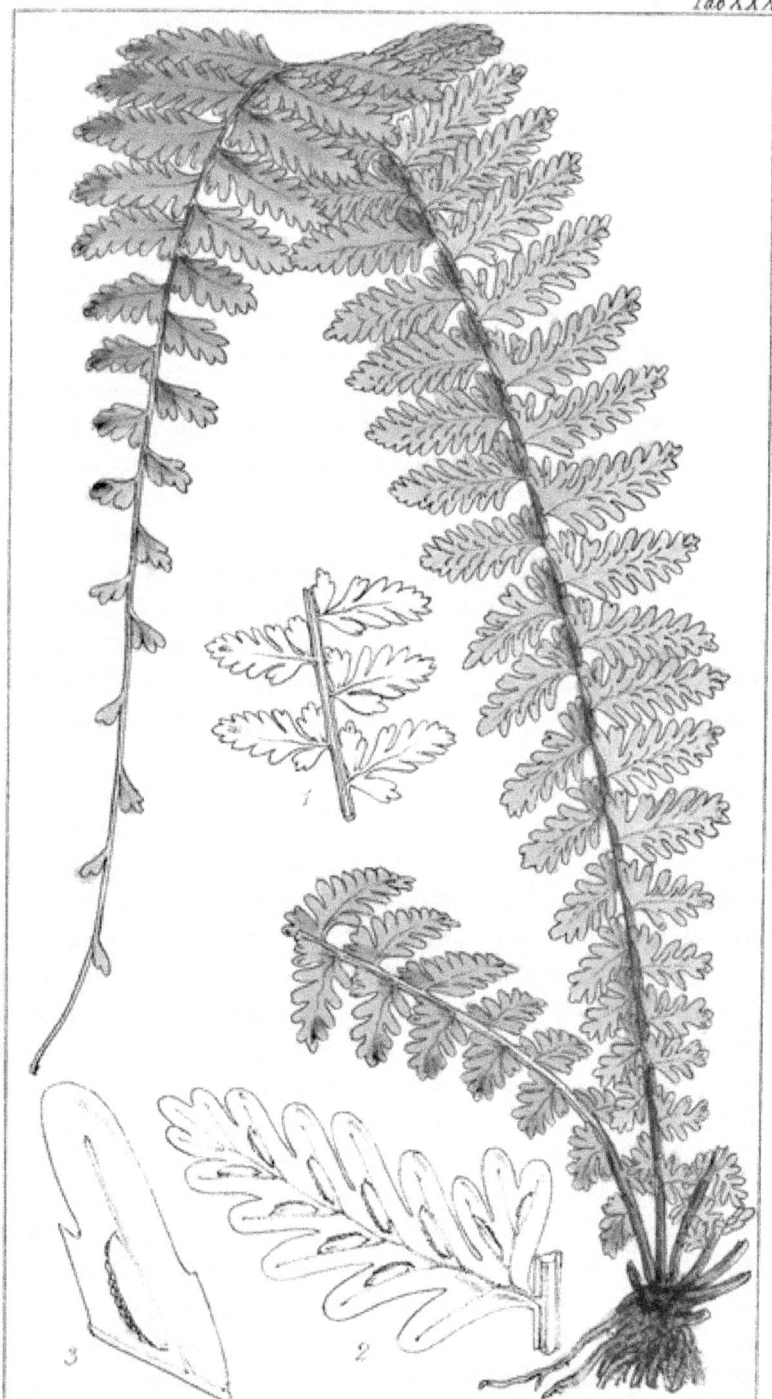

TAB. XXXI.

ASPLENIUM (EUASPLENIUM) REPENS, *Hook.*

Parvum, caudice longe filiformi ramoso hirsuto-tomentoso, stipitibus sparsis remotis vix 2 lineas longis, frondibus sub-biuncialibus ovato-lanceolatis bi-rarius tripinnatis, pinnis 2 lineas longis petiolatis, pinnulis divaricato-patentibus vix lineam longis cuneato-palmatis in petiolulum attenuatis apice irregulariter subdigitato-laciniatis, venis crassiusculis immersis simplicibus vel furcatis longe ante apicem terminantibus clavatis, soris semiovatis solitariis, involucris membranaceis fuscis, rachi stipiteque herbaceis.

Asplenium repens, *Hook. Sp. Fil.* 3. *p.* 194.

HAB. Ecuador, growing on trees and shrubs in the forests of Archedona, Quitinian Andes, *Jameson, n.* 786.

One of the most distinct of all *Asplenia.* The filiform caudices are a foot and more long; yet the fronds are among the smallest of the Genus, and the pinnules of the fronds are in shape more like some *Plagiochilus* among *Jungermannia* than any Fern I know. The veins terminate far below the apex of the segment, and the involucres are all remote from the margin.

Portion of a fertile plant: *nat. size. Fig.* 1 & 2. Pinnæ, with and without sori. *f.* 3. sorus: *magnified.*

Tab XXXIV

TAB. XXXII.

ONYCHIUM STRICTUM, *Kze.*

Caudice subfusiformi tuberculato obliquo apice radicoso, stipitibus cæspitosis spithamæis ad pedalem rachibusque stramineis inferne parce paleaceis, frondibus subspithamæis fere
membranaceis viridibus glabris opacis subdeltoideo-ovatis
3-4-pinnatis seu pinnatisectis, segmentis lineari-subcuneatis
acutis sæpe oppositis integris v. bitrifidis, rachibus anguste
alatis, fertilibus paulo latioribus, soris brevibus sub-oblongis
obtusis curvatis ante apicem sitis rarius solitariis.

Onychium strictum, *Kunze, in Schk. Fil. Suppl.* 2. *p.* 11,
(*no figure*). *Hook. Sp. Fil.* 2. *p.* 123.

HAB.   St. Jago de Cuba, on Mount Leban, *Linden, n.* 1870,
C. *Wright, n.* 1858.

With the exception of the present species, and *Onychium
angustifolium* Kze. (our *Pellæa decomposita,* v. 2. p. 171) all
the true *Onychia* are natives of the Old World.   Here however it may be observed that the sori are less decidedly in
exactly opposite pairs than is consistent with the character
of the Genus.   It borders too closely on *Cheilanthes,* as that
does again on *Adiantum.*

   Fertile plant : *nat. size.   Fig.* 1. Portion of a pinna with
sori : *magnified.*

Tab. XXXII.

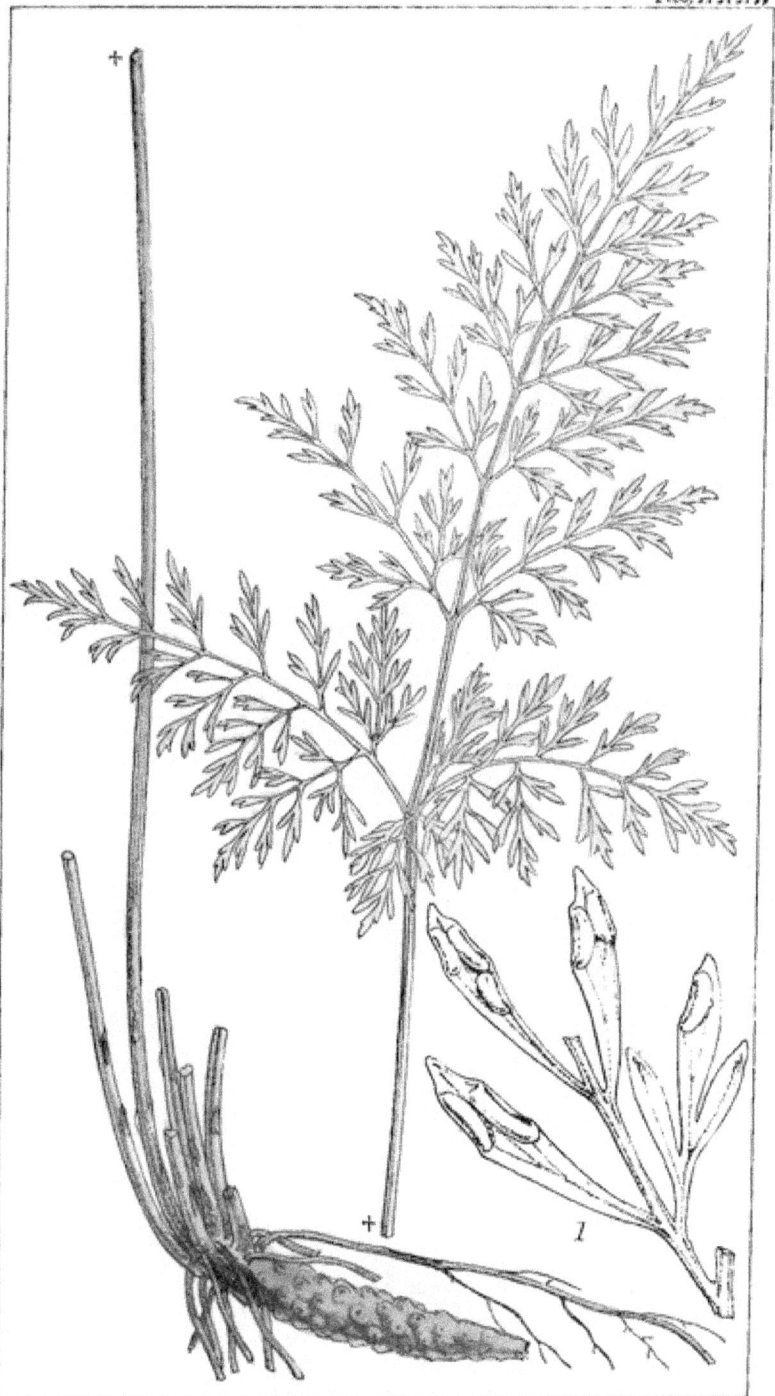

TAB. XXXIII.

Asplenium (Euasplenium) Wardii, *Hook.*

Caudice subhorizontali crassiusculo superne squamis longis
subulatis dense vestito, stipitibus cæspitosis spithamæis
stramineis inferne parce subulato-squamosis, frondibus 12-
14 uncias longis basi 10 uncias latis membranaceis deltoideo-
acuminatis olivaceo-fuscis opacis bipinnatis apice pinnatifidis, pinnis horizontalibus petiolatis lanceolatis acuminatis,
pinnulis approximatis 6-8 lineas longis horizontalibus sessilibus subdimidiato-ovatis obtusis integerrimis v. læviter
sinuatis serratisve, pinnarum inferiorum pinnulis subpinna-
tifidis auriculatisque superiorum subintegris decurrentibus,
venis pinnatis oblique patentibus simplicibus furcatisve,
soris biserialibus costam versus, involucris (junioribus)
tenui-membranaceis pallidis.

Asplenium Wardii, *Hook. Sp. Fil.* 3. *p.* 189.

Hab. Tsus Sima, Strait of Korea, *Wilford, n.* 717.

A very distinct *Asplenium* from any known to me, with
more the habit of some *Lastrea* than is usually seen in this
Genus. I name it in compliment to John Ward, Esq. Com-
mander of H. M. S. "Actæon," in acknowledgement of his
services rendered to Mr. Wilford, Botanical Collector for the
Royal Gardens of Kew, during an interesting cruise in the
North Chinese Seas, and as far as Manchuria.

Fertile plant: *nat. size. Fig.* 1. Pinnule with sori: *magnified.*

Tab. XXXIII

TAB. XXXIV.

ASPLENIUM (EUASPLENIUM) RUTACEUM, *Mett.*

Caudice obliquo radicante, stipitibus cæspitosis brevissimis
semiunciam ad duas uncias longis castaneis, frondibus 10-12
uncias longis membranaceis atro-viridibus lato-lanceolatis
basi sensim attenuatis apice in caudam longam filiformem
ad extremitatem radicantem extensis bi-tripinnatis, pinnis
primariis horizontalibus unciam sesquiunciam longis numerosis approximatis (infimis nanis) e basi latiuscula oblongis
obtusis, secundariis 2 lineas longis omnibus petiolatis pinnatis, pinnulis 2-3 obovato-spathulatis subacutis integris
v. bilobis, pinnulis infimis magis compositis summis integris
vel bilobis minimis remotis, venis in quoque lobo indivisis
longe infra apicem terminantibus apice clavatis, soris brevibus ovalibus in disco sitis, involucris membranaceis pallidis.

Asplenium rutaceum, *Metten. Asplen. p.* 129. *t.* 5. *f.* 32. 33.
Moore. Ind. Fil. p. 162. *Hook. Sp. Fil. p.* 203. Aspidium,
Willd. Sp. Pl. 5. *p.* 266. Athyrium, *Pr.*—Lonchitis in
auriculas subrotundas divisa. *Plum. Fil. p.* 44. *t.* 57.

HAB. St. Domingo, *Plumier;* Columbia, Tovar, *Moritz, n.*
402; New Grenada, Ocaña, *Schlim n.* 624; and Sierra
Nevada, elev. 6000 feet; Venezuela, *Fendler, n.* 123; on
trunks of trees, forests of Archedona, Andes of Quito,
Jameson, n. 788.

An elegant species, lately well described by Mettenius,
previously very incorrectly understood, and chiefly in consequence of Plumier's rather exaggerated figure above quoted,
from which Willdenow's character appears to have been
drawn up : and hence too he was led into the error of
believing it to be an *Aspidium.*

Fertile plant: *nat. size. Figs.* 1 & 2. Pinnules with sori;
magnified.

Tab XXXIV

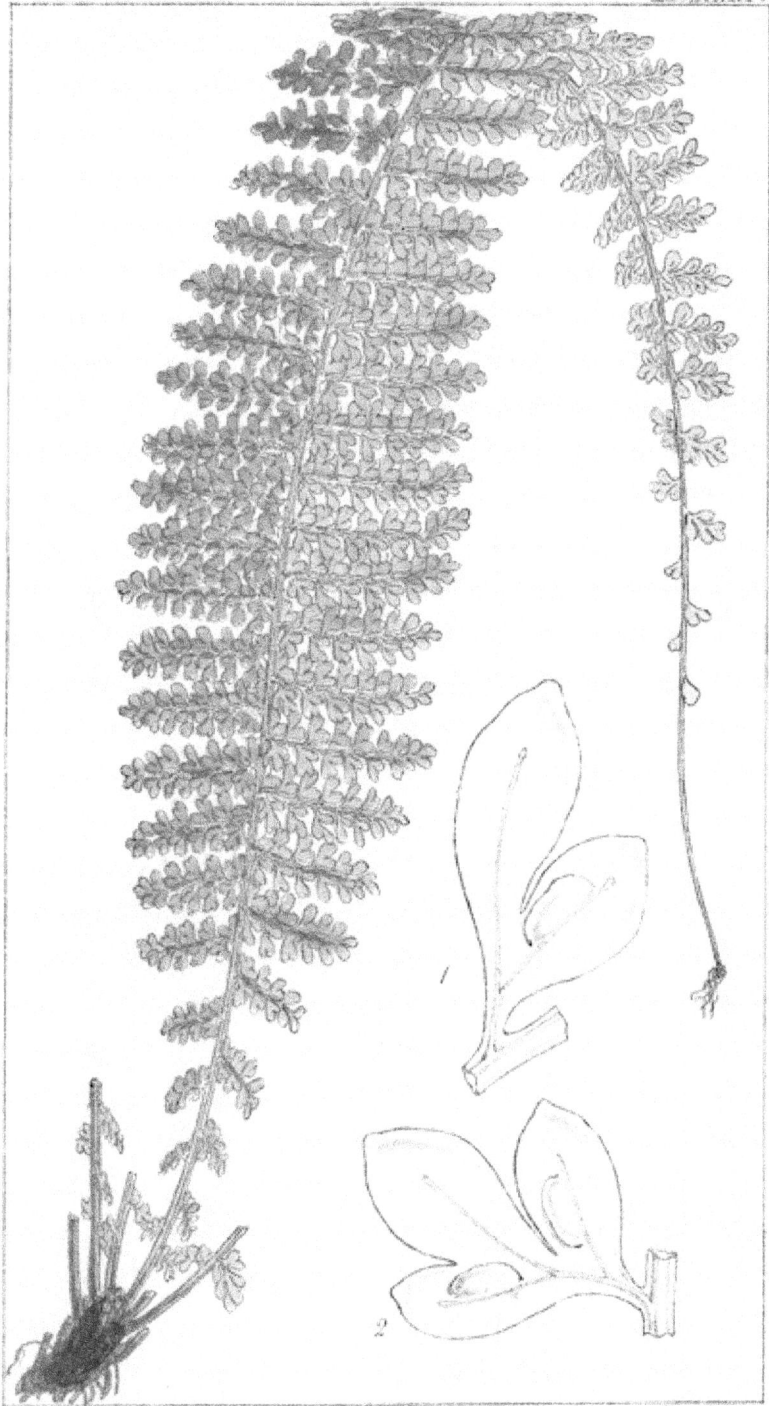

TAB. XXXV.

SCOLOPENDRIUM (CAMPTOSORUS) SIBIRICUM, *Hook.*

Caudice parvo adscendente radiculoso, stipitibus cæspitosis gracilibus 2-4-uncias longis, frondibus membranaceis, *sterilibus* brevibus oblongo-ovatis acuminatis, *fertilibus* 5-6 uncialibus lanceolatis longissime caudatim attenuatis apice radicantibus, venis prope costam anastomosantibus reliquis liberis apicibus clavatis, soris geminatis non raro sparsis solitariis.

Scolopendrium (Camptosorus) sibiricum, *Hook. Sp. Fil.* 3. *ined.*

Camptosorus Sibiricus " *Ruprecht in Beitr.* 2, *Pflanzenk. d. Russ. R. III. p.* 45." *Ledeb. Fl. Ross.* 4. *p.* 523.

HAB. Siberia, River Angara, *Steller;* Kamtschatka, *Georgi,* Island of Tsus Sima, Strait of Korea, *Wilford, n.* 790.

Linnæus gives "Siberia" as a locality for the N. American *Asplenium* (Camptosorus) *rhizophyllum.* The Siberian plant is however since acknowledged to be a new species, distinguished by the entire absence of lobes or auricles at the base of the frond, which are so characteristic of the United States. But even in Siberia the present species appears to be of very rare occurrence, insomuch that the late learned author of the Flora Rossica (Ledebour) was obliged to declare "species mihi ignota." Only two stations for it have been yet recorded in all the Russian dominions; and now that it has been detected in the island of Tsus Sima, off the coast of Korea, the discoverer there accompanies his specimen by the remark " the only specimen found;" and that is the one here represented.

Camptosorus and *Antigramme* only differ from *Scolopendrium* by the partial anastomosing of the veins; in the former next the costa; in the latter next the margin. To me it seems most natural to unite both with *Scolopendrium.*

Fertile and barren fronds. *Fig.* 1. Portion of a sterile frond, showing the venation, and *f.* 2. portion of a fertile frond with sori : *magnified.*

Tab. XXXV

TAB. XXXVI.

ASPLENIUM (EUASPLENIUM) DIMORPHUM, *Kze.*

Caudice ("repente crasso") stipitibus 6-12 uncias longis, frondis amplis 2-pedalibus et ultra chartacco-membranaceis deltoideo-ovatis bi-tripinnatis, pinnis ovato-lanceolatis petiolatis acuminatis biformibus in eadem fronde v. in frondibus diversis, inferioribus plerumque sterilibus simpliciter pinnatis, pinnulis rhombeo-ovatis basi inequaliter cuneatis margine serratis sæpe lobatis vel basin versus iterum pinnatis, venis pinnatis dichotomis ; pinnis fertilibus plerumque terminalibus bipinnatis vel potius bipinnatifidis, laciniis linearibus angustis obtusis, venis costiformibus, soris plerumque solitariis dareiformibus ramis geminatis diplazioideis.

Asplenium dimorphum, *Kze, in Linnæa*, 23. *p.* 233. *Metten. Asplen. p.* 108. (*excl. syn.* A. Novæ Caledoniæ, *Hook.) Hook. Sp. Fil.* 3. *p.* 213. Aspl diversifolium. *A Cunn. in Endl. Fl. Norf. p.* 10. (*not of Blume.*)

HAB. Norfolk Island, and no where else as far as yet known.

One of the most distinct of asplenioid Ferns, and among the most limited in respect of country, for it appears to be confined to Norfolk Island.

Fig. 1. Portion of a frond, with sterile and fertile pinnæ; *nat. size. f.* 2. Sterile pinnule. *f.* 3. Fertile pinnule with sori : *magnified.*

Tab. XXXVI

TAB. XXXVII.

ASPLENIUM (EUASPLENIUM) SCANDENS, *J. Sm.*

Caudice crassitie pennæ corvinæ longe repente flexuoso ramoso parce radicante apicibus paleaceis, stipitibus sparsis
remotis brevissimis, frondibus 1-2-pedalibus et ultra late
ovato-lanceolatis basin versus attenuatis subcoriaceo-membranaceis olivaceo-viridibus 3-4-pinnatis, pinnis primariis
horizontalibus 3-4-uncias longis lato-lanceolatis sessilibus
numerosis subdistantibus, inferioribus nanis magis remotis,
pinnis secundariis unciam longis, pinnulis ultimis seu laciniis 4-5-lineas longis anguste linearibus infimis superioribus furcatis vel trifidis reliquis integris acutis, fertilibus
paululum latioribus, venis costæformibus, soris marginalibus
oblongis, involucris firmis submembranaceis pallide fuscis;
rachibus primariis teretibus, partialibus compressis sub-
alatis.

Asplenium scandens, *J. Sm. in Hook. Journ. of Bot.* 3. *p.*
408. *(name only).* *Metten. Asplen. p.* 108. *Hook. Sp. Fil.*
3. *p.* 216.

HAB. Philippine Islands, Leyte, *Cuming. n.* 297; New
Guinea, *Hinds.*

A rare and well-marked species, remarkable for its long
creeping, or perhaps, scandent caudex.

Fig. 1. Caudex and small frond. *f.* 2. Portion of a larger
and fertile frond; *nat. size. f.* 3. Sorus: *magnified.*

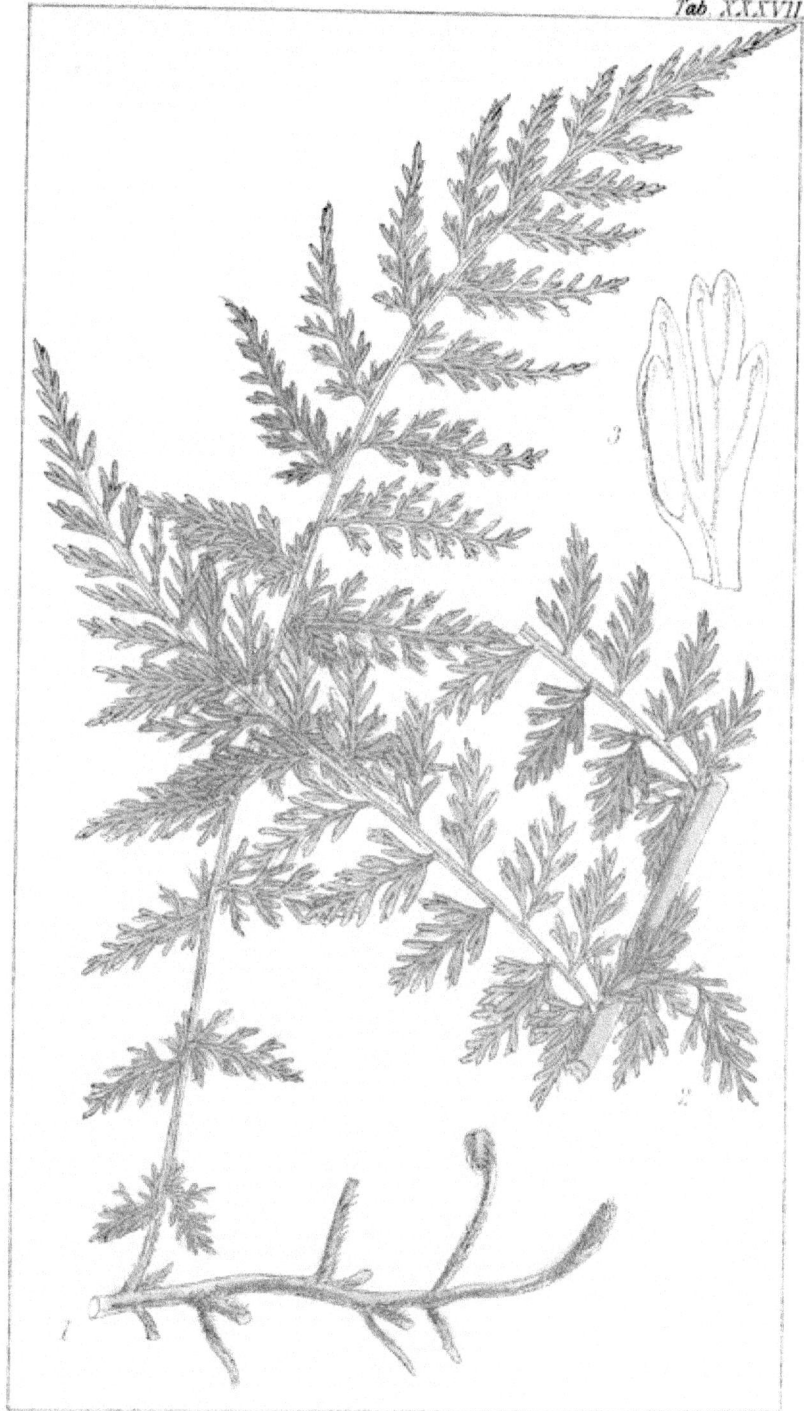

Tab. XXXVII

3

2

1

TAB. XXXVIII.

Asplenium (Euasplenium) ferulaceum, *Moore*.

Caudice? stipite 14 uncias longo robusto pallide fusco hinc
sulcato, fronde sesquipedali supradecomposita (4-5-pinnata)
læte viridi membranacea deltoideo-ovata acuminata, pinnis
primariis numerosis subdistantibus inferioribus 6-8 uncias
longis petiolatis late ovatis acuminatis, secundariis 3 uncias
longis, ultimis brevibus lineam longis lineari-subspathulatis
mono-rarius disoris, venis costæformibus, soris parvis darc-
oideis, involucris viridescentibus submembranaceis, rachibus
primariis secundariisque teretibus stramineis nitidis ultimis
augustissimis compressis glabris.

Asplenium ferulaceum, *Hook Sp Fil.* 3. *p.* 216. *Moore. Mss.
in Herb. Nostr. Moore, Ind. Fil. p.* 130. *(name only, no
character or description.)*

HAB. New Grenada, *Hartweg. n.* 1519. Quito, *Jameson, in
Herb. Nostr.*

A very distinct and elegant, and hitherto undescribed
species, of the *Darea*-group, remarkable for the very compound, or rather decompound, finely cut pinnæ, and the
terete primary and secondary rachises, which are stramineous
and glossy.

Fig. 1. Stipes and base of a lower primary pinna. *f.* 2. A
superior primary pinna; *nat. size. f.* 3. Pinnule with sori:
magnified.

Tab. XXXVIII

TAB. XXXIX.

ASPLENIUM (EUASPLENIUM) DICHOTOMUM, *Hook.*

Parvum, caudice erecto subnullo radicante paleaceo, stipitibus cæspitosis subunciam longis gracilibus compressis pallide viridibus setaceo-palcaceis, frondibus 3-4-uncialibus oblongis acutis membranaceis viridibus subtripinnatis, pinnis primariis ½-¾ unciam longis subdimidiato-ovatis subapproximatis petiolatis dichotome divisis, pinnulis brevibus augustis linearibus bis-terve dichotomis, segmentis fertilibus paulo latioribus, venis solitariis costiformibus, soris magnis lineari-oblongis marginalibus dareoideis, rachibus omnibus compresso-alatis.

Asplenium dichotomum, *Hook. Sp. Fil. 3. p.* 210.

HAB. Borneo, *Hugh Low, Jun., Esq.;* Lobouk Peak. elev. 5,000 feet, north-east side of Borneo, *Thos. Lobb.*

This would be a true *Darea* (or *Ceratopteris,*) in the view of those botanists who adopt that genus; but I know none with which it is likely to be confounded. It is a small and extremely delicate species, the whole height scarcely exceeding 4 inches.

Fertile plant; *nat. size. Fig.* 1. Pinna with sori. *f.* 2. Single sorus : *magnified.*

CENT. 2. T. 39.

Tab. XXXIX.

TAB. XL.

ASPLENIUM (EUASPLENIUM) DAVALLIOIDES, *Hook.*

Caudice parvo suberecto supra squamoso, stipitibus cæspitosis 3-4 uncias longis compressis deciduo squamosis, frondibus 3-6-uncias longis subcoriaceis ovatis acuminatis 3-subquadripinnatis (junioribus primariis pinnis lobato-pinnatifidis), pinnis primariis 2-3 uncias longis petiolatis, secundariis petiolatis ¼-¾ unciam longis late ovatis, pinnulis ultimis parvis oblongis acutis simplicibus vel bifidis, segmentis omnibus patenti-recurris acutis, venis costiformibus, soris oblongis sæpissime marginalibus longitudine fere segmentorum, involucris membranaceis firmis, rachibus alato-compressis.

Asplenium davallioides, *Hook. Florul. Hong-Kong. in Kew Gard. Misc. 9. p.* 343. *Hook. Sp. Fil. p.* 212.

HAB. Nangasaki, Japan, *Babington, n.* 101; Loochoo Islands, *C. Wright;* Tsus Sima, Strait of Korea, *Wilford, n.* 791.

A species which cannot easily be mistaken for any other of the group to which it belongs, if the short and singularly divergent or patenti-recurved ultimate segments be considered, with the sori, though short in themselves, nearly as long as the ultimate segments, and giving the appearance of a dareoid species of *Davallia.*

Fertile plant; *nat. size. Fig.* 1. Pinnule with sori. *f.* 2. 3. Back and front view of a sorus; *magnified. f.* 4. Young frond ; *nat. size.*

Tab. XL.

TAB. LXIV.

Davallia (Cuneatæ?) trichomanoides, *Hook.*

Caudice? stipite subspithamæo (et ultra?) fusco-viridi terete, fronde perelegante pedali et ultra ovata acuminata 4-pinnata seu decomposita, pinnis primariis 4-8 uncias longis ovato-lanceolatis, secundariis 1-2 uncias longis, laciniis omnibus angustissimis lineari-spathulatis obtusis vix lineam longis plerisque soriferis, soris terminalibus, involucro oblongo-lingulato segmentis angustiore, costa lata, rachibus stramineo-viridibus nitidis.

Hab. Wooded Mountains, Naviti Levu, Fiji Islands, *Alex. Milne* in Voyage of H.M.S. Herald, Capt. Denham, R.N.

This is assuredly the most elegant of the genus *Davallia*, so narrow in its segments that they seem to be composed of the midrib with a very narrow herbaceous margin. The apices of the fertile segments which are very copious have the appearance of being unequally 2-lipped: the segment itself being much dilated, and the involucre, an oblong or tongue-shaped scale, is appressed to it, quite concealing the sorus or cluster of capsules in its axis.

Tab. LXIV. Represents a primary pinna of Davallia trichomanoides, with fructification; *natural size. Fig.* 1. Segments, of which one is fertile; *magnified. f.* 2. Apex of a segment with involucre, and *f.* 3, the same, with the involucre removed, showing the sorus of capsules on long pedicels: more *magnified.*

CENT. 2. T. 64.

LXV,

TAB. LVII.

Aspidium (Polystichum) tripteron, *Kze.*

Caudice brevi erecto crasso squamis magnis ovatis acuminatis paleaceis, stipitibus cæspitosis semipedalibus ad pedalem stramineis inferne squamosis, frondibus 1-1½ ped. submembranaceis circumscriptione hastato-lanceolatis acuminatis pinnatis, pinnis numerosis 1½ unciam longis horizontalibus sessilibus e basi oblique cuneata superne auriculata lanceolatis grosse serratis setoso-mucronatis subtus rachibusque parce albo-paleaceis, infimis duabus suboppositis 4-6-uncialibus iterum pinnatis, soris plerisque subbiserialibus, involucris parvis orbicularibus peltatis demum obsoletis.

Aspidium tripteron, *Kze. Bot. Zeit.* 6. p. 509; *Metten. Aspid.* p. 51.

Hab. Island of Tsus Sima, Gulf of Korea, *Wilford.* Japan, *Goring*; Hakodadi, *C. P. Hodgson, Esq.*

A species well distinguished among the *Polystichum*-group of *Aspidium* by the flaccid and submembranaceous texture, simply pinnate, except the lowest pair of pinnæ; these are greatly elongated and again pinnate, resembling two ears, and spreading so horizontally as to form in circumscription a hastate frond. In the more advanced state of the fructification, the delicate involucre is so shrunk and concealed by the copious capsules of the sori that the Fern resembles a *Polypodium* (§ Phegopteris.)

Tab. LVII. Aspidium tripteron; *nat. size. Fig.* 1. Portion of a pinna with young sori. *f.* 2. An old sorus, with the involucre nearly obsolete;—*magnified.*

Tab.LVII

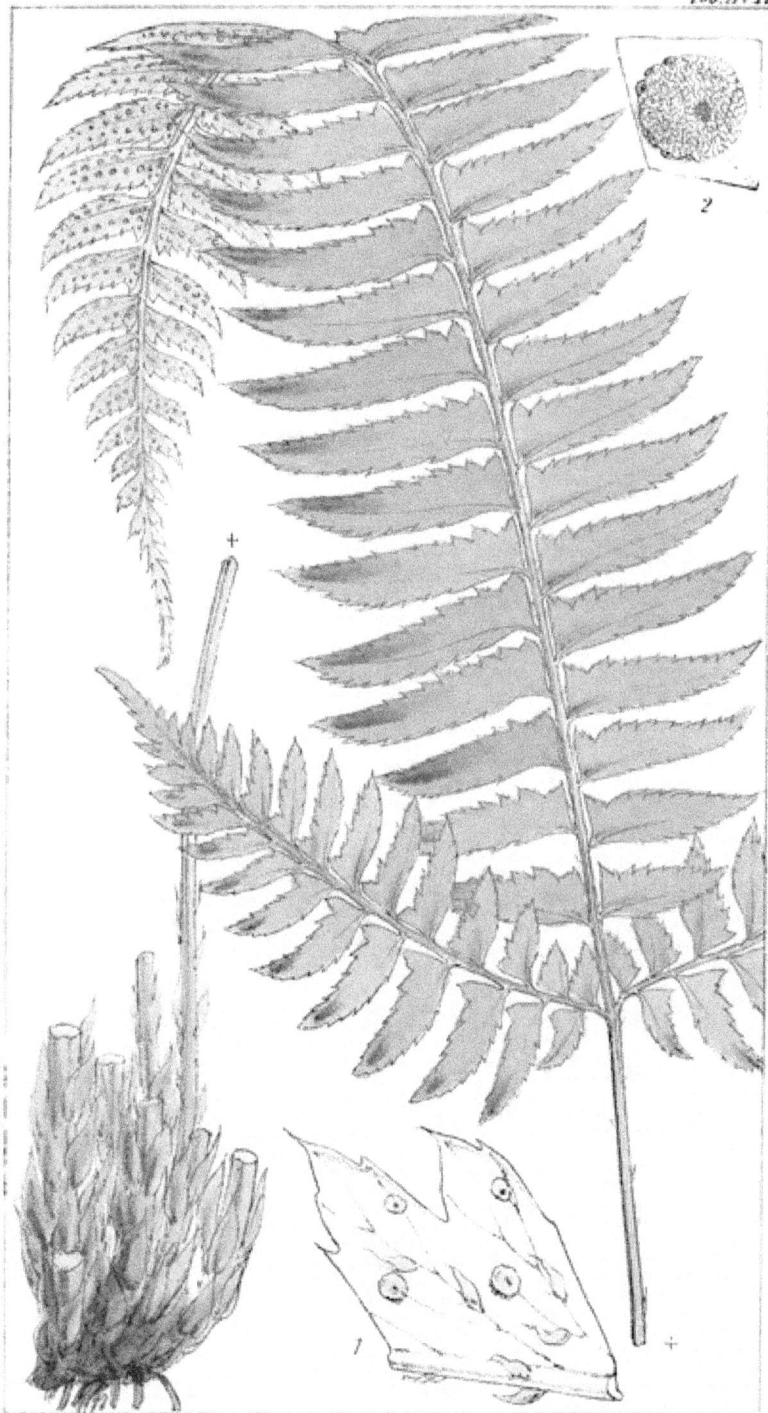

1

2

TAB. LVIII.

NIPHOBOLUS LINEARIFOLIUS, *Hook.*

Caudice longe repente ramoso sæpe copiose radiculoso
squamis subulatis ferrugineis dense imbricatis nitidis
setoso-paleaceo, frondibus sparsis erectis carnoso-coriaceis
sessilibus 3-4 uncias longis 1 lineam latis linearibus totis
pilis stellatis pallide ferrugineis tectis, demum superne præcipue nudiusculis viridibus, soris oblongo-rotundatis biserialibus series in dimidiam superiorem frondis, venis remotis
anastomosantibus, areolis appendiculatis.

HAB. Island of Tsus Sima, gulf of Korea, growing on rocks
along with *Pleopeltis nuda*, Hook. (Polypodium sesquipe-
dale, *Wall.*)

The genus *Niphobolus* of Kaulfuss, is one among the *Poly-
podioid* Ferns that is retained by some Botanists and rejected
by others. Few as are its characters, it is generally easily
distinguished by its habit and its stellato-tomentose covering.
Its venation is very indistinct and difficult to be seen and to
be accurately represented, and is said by Presl to be different
in different species. Then as to specific differences they de-
pend mostly on the outline of the fronds which are uniformly
undivided. In favor of this being considered distinct I may
observe that its fronds are so narrow and so uniformly small
(4 inches being the extreme length) that it is at once distinguishable from every other kind known to me.

TAB. LVIII. Plant of Niphobolus linearifolius; *nat. size.*
Fig. 1. Apex of a fertile frond;—*magnified.* *f.* 2. Fertile
portion, more highly *magnified*, and exhibiting the venation.
f. 3. Tuft of stellated hairs; very highly *magnified.*

Tab. LVIII.

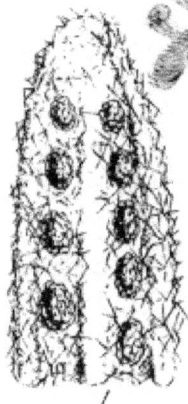

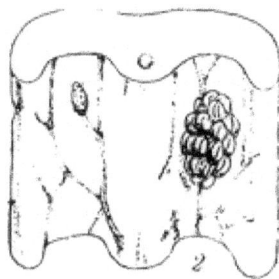

TAB. LIX.

HYPOLEPIS PTERIDIOIDES, *Hook.*

Stipite stramineo, fronde (ut videtur) ampla, basi trichotome
et pedatim divisa 3-4-pinnata submembranacea glabra siccitate olivacea, pinnis primariis subsesquipedalibus oblique
ovatis acuminatis petiolatis reliquis sessilibus lanceolatis
acuminatis profunde fere ad costam pinnatifidis apicibus
acuminatis serrato-lobatis, laciniis vix semipollicaribus oblongis obtuse crenato-serratis margine utrinque monosoris, venis
remotis liberis infra medium furcatis ramis patentibus, soris
(in depressione seu cavitate frondis) impressis, involucris
transversim oblongis fuscis, rachibus stramineis nitidis, costis
versum apicem pinnarum supra spinulosis.

HAB. Peak of Fernando Po, at 7000 feet elevation, *Gustav
Mann.* n. 348.

The habit of this Fern, whose fronds are probably too
large to be preserved entire, seems to be quite that of some
exotic species of *Pteris* (Eupteris) of the group to which
Pt. arguta, Ait. and *Pt. flabellata,* Thunb. belong, and in
some of them the sori are in a degree abbreviated, but here
they are as much so as (or more than) in many species of
Adiantum, and which have induced me to refer this species
to *Hypolepis.* In the present plant I never find more than
one sorus on each side of a segment, and these sori are invariably sunk in a cavity or depression, which occasions a cor-
respondent swelling, of the same form, on the superior side
of the frond.

TAB. LIX. *Fig.* 1, 2. Portions of a fertile frond of Hypolepis pteridioides;—*nat. size. f.* 2, 3. Sori seen from beneath, and *f.* 4. Impressions caused by the sori, as seen on
the superior side of the segment; more or less *magnified.*

Tab. LIX.

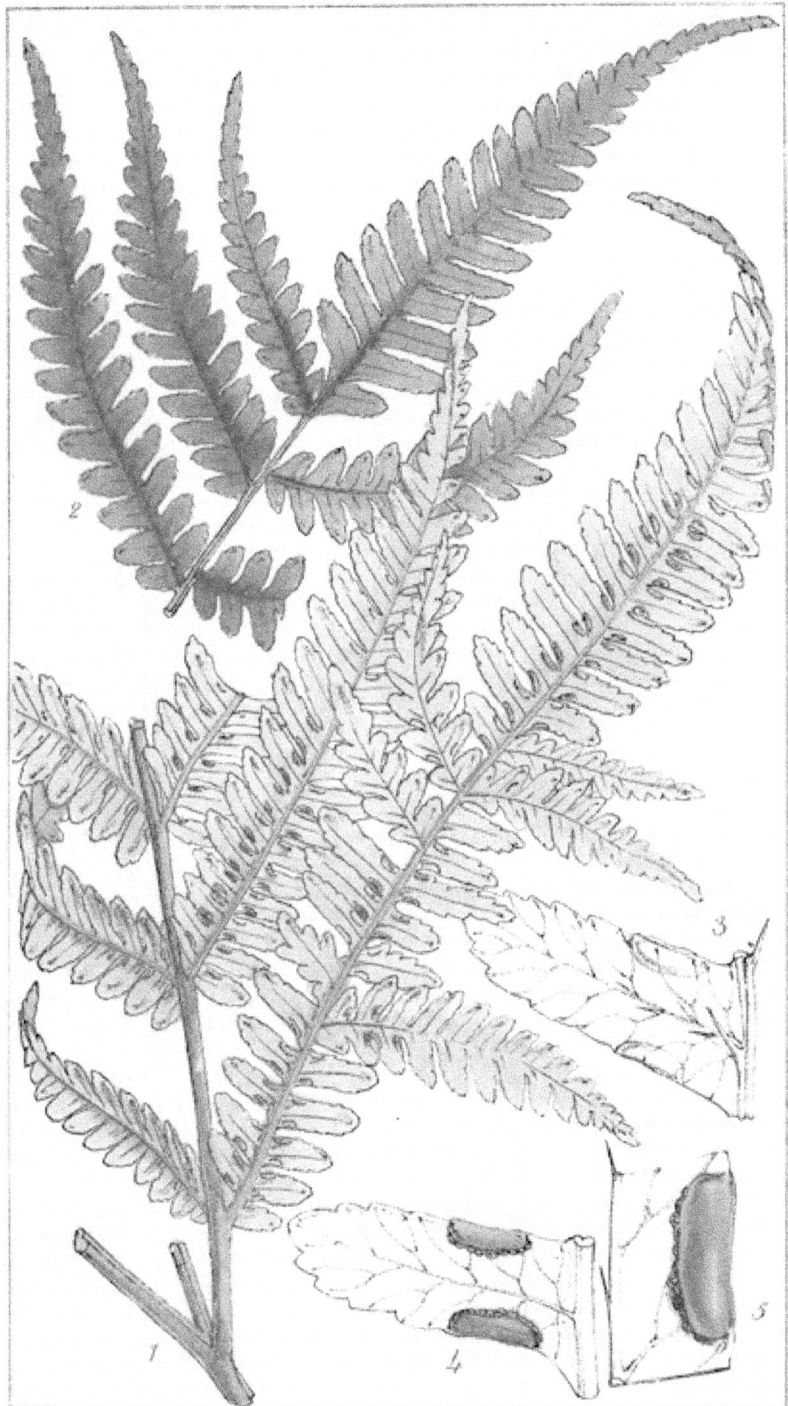

TAB. LX.

ASPLENIUM (DAREA) MANNII, *Hook.*

Nanum, caudice filiformi sarmentoso longe repente intricato
copiose radiculoso, stipitibus subaggregatis vix biuncialibus
gracilibus, frondibus biuncialibus oblongo-ovatis subcarnoso-
coriaceis viridibus bipinnatis, pinnis remotis, pinnulis subu-
nilateralibus, sterilibus linearibus obtusis, fertilibus apice
oblique semiovatis rostratis monosoris, involucro (ratione
plantæ) amplo laxe membranaceo.

HAB. Peak of Fernando Po, elev. 3000 feet above the level
of the sea, epiphytal, *Gustav Mann.*

Evidently one of the *Darea*-group of *Asplenium*, but extremely unlike any described species, though allied to *A.
brachypterum* of Kunze : yet it can hardly be a state of that ;
the caudex and pinnules and habit are so different.   The former resembles a slender filiform stolon, forming dense copious
intricated masses, at distances throwing out tufted fibrous
roots, on the undersides, and a few clustered fronds on the
upper.   The fertile segments with their oblique semiovate
apex almost exactly resemble in shape the *Buxbaumia aphylla*
among mosses, and the delicate membranaceous involucre is
large and lax.   The term pinnatifid is more applicable to
these fronds than pinnate : a single vein passes through the
centre and terminates in a clavate apex below the extremity
of each segment; from the upper side this sorus arises.

TAB. LX. Portion of a tuft of Asplenium Mannii, *nat. size.*
Fig. 1. Anterior side of a fertile pinna or segment, and *f.* 2.
Posterior, or underside of ditto ;—*magnified.*

TAB. LXI.

ADIANTUM (EUADIANTUM) FLEXUOSUM, *Hook.*

Caudice ascendente nodoso-ramoso dense squamis subulatis aterrimis paleaceo, stipitibus subaggregatis erectis ebeneis piloso-asperis, frondibus sesquipedalibus et ultra oblongo-lanceolatis seu ovatis coriaceo-membranaceis atro-viridibus 3-4-pinnatis, pinnis primariis magis minusve refractis oblongis inferne bipinnatis, pinnulis petiolatis reniformi-obcordatis, sterilibus varie lobatis, fertilibus margine integris, venis flabellatim dichotomis, soris approximatis subuniformibus elliptico-oblongis coriaceis, rachibus ubique insigniter angulato-flexuosis pubescenti-tomentosis scabriusculis.

HAB. Santa Rosa, Vera Paz, in hollows of pine ridges, Guatemala, *Osbert Salvyn, Esq.*

A new species and a very distinctly marked one, peculiar in the scandent habit, in the refracted pinnæ (as is common in other scandent ferns) and in the asperous zigzag rachis.

TAB. LXI. *Fig.* 1, 2, 3. Portions of Adiantum flexuosum; *nat. size. f.* 4. Fertile pinnule; *magnified. f.* 5. Sorus, more highly *magnified.*

CENT. T. I. 61.

Tab LXI

TAB. LXII.

NEPHRODIUM (LASTREA) MILNEI, *Hook.*

Stipite rachibusque primariis intense nigro-ebeneo nitidis,
frondibus sesquipedalibus ovatis accuminatis membranaceis
atro-viridibus opacis bipinnatis, pinnis primariis ovatis
petiolatis, secundariis sessilibus lanceolatis profunde pinnatifidis lobis oblongis obtusiusculis sinuato-dentatis sub-pin-
natifidisve, sinubus lobulo acuto donatis, venis remotis simplicibus medium versus unisoris, involucro carnoso-celluloso
cordato-reniformi, marginibus dentatis dentibus glandula
globosa terminatis.

Hab. Wooded mountains, interior of Naviti Levu, Fiji
Islands, *Alex. Milne*, Denham's Voyage of H.M.S. Herald.

I cannot refer this to any described species of the *Lastrea*
group of *Nephrodium*. It presents few striking distinguish-
ing characters. The stipes (as much as was gathered) and
the main rachises are indeed singularly black and ebeneous;
and on many of the ultimate pinnæ the sinuses are furnished
with an acute lobe, or tooth (see our *fig.* 2).

TAB. LXII. *Fig.* 1. Portion of Nephrodium Milnei;—
nat. size. f. 2. Segment of a pinna with sori;—*magnified.
f.* 3. Sorus;—more highly *magnified.*

Tab LXIV.

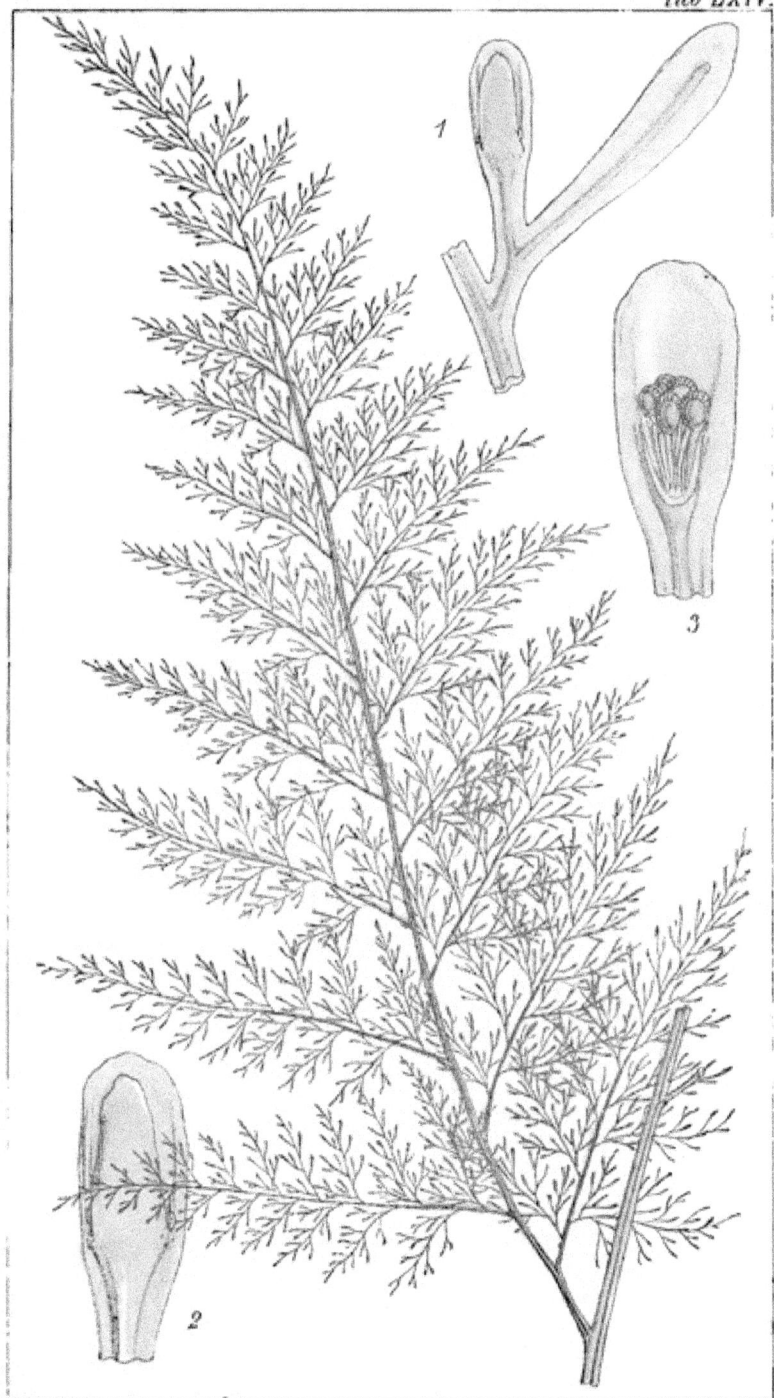

TAB. LXV.

ANEMIA MEXICANA, *Kl.*; *var.* paucifolia.

Elata, frondibus pinnatis, pinnis 3-13 ovatis obtusis magis
minusve acuminatis serratis brevi-petiolatis glabris penni-
veniis, venulis dichotomis liberis, spicis pedunculatis pinnatis, pinnis pinnatifidis apice confluentibus lobis capsuliferis oblongis longe villosis.

Anemia Mexicana, *Kl. in Linnæa*, 18, p. 526, *Kze. in Schk.*
Fil. Suppl. p. 75, t. 131. *Hook. Icon. Plant* t. 988.

β. *paucifolia;* minor, frondibus ternatis. (Tab. Nostr. LXV.)

Anemia speciosa, *Presl, Suppl. Pterid.* p. 89; *Liebm. Fil. Mex.*
p. 151.

HAB. β. Western Mexico, *Neé.* Mountains of Oaxaca, elev.
4500 feet, *Liebmann.* Lofty mountains of Guatemala, Lan-
quin, Vera Paz, elev. 2000 feet, *Osbert Salvin, Esq.*

This plant which is maintained as a distinct species, under
the not very appropriate name of *A. speciosa*, by Presl and
Liebmann (from whom I have received authentic specimens),
I can only consider as variety of *Anemia Mexicana*, a variety
probably occasioned by its locality, at a considerable elevation
on the mountains. The fronds are of a firmer texture, the
pinnæ are fewer, generally 3, and these shorter and blunter.

TAB. LXV. Plants, sterile and fertile, of *Anemia Mexicana*,
β.; *natural size.* *Fig.* 1. Portion of a sterile pinna: and *f.*
2, Portion of a fertile spike; *magnified.*

Tab LXV

TAB. LXVI.

ALSOPHILA PODOPHYLLA, *Hook.*

Frondibus bi-tri-(quadri?)-pinnatis glabris, pinnulis 4-6-polli-
caribus petiolatis lineari-oblongo-lanceolatis coriaceo-membranaceis acuminatis sinuato-dentatis basi truncatis apice
serratis terminali pinnatifido-lobato majore subtus pallidioribus, venis infimis solummodo cum proxima vena anastomosantibus, soris copiosis sparsis globosis, capsulis com-
pactis, receptaculo magno hemisphærico, costa subtus pu-
bescenti-squamulosa.

Alsophila podophylla, *Hook. in Kew Gard. Misc.* 9, p. 334.

HAB. Chusan, *Alexander.* Hong Kong, *Dr. Harland, J. C.
Bowring, Esq. Col. Urquhart*; abundant at the foot of Vic-
toria Peak, in a ravine, *Wilford.*

A very distinct and well marked species; its nearest ally is
probably *Als. gigantea, Wall.* The caudex is 4-8 feet high.
Fronds 8-9 feet long. Stipes densely scaly below, and, as
well as the main rachises, bright castaneous when dry, rough
to the touch, but not to the naked eye, with minute raised
points. Veins in fascicles; the lowest veinlet in each fascicle
very frequently uniting with the lower one of the opposite
fascicle and thus forming a triangular areole next the costa,
the rest of the veinlets free. This plant would probably be
a *Gymnosphæra* of Blume and may be, and possibly is, his *G.
glabra*; but with the very brief character given by that
author of less than two lines, it is impossible to form any
decided opinion, one way or the other.

TAB. LXVI. *Figs.* 1, 2, 3. Portions of a frond of *Alsophila podophylla*, sterile and fertile; *natural size. f.* 4. Portion of a fertile pinna, showing the venation; *magnified. f.* 5.
Single sorus, and *f.* 6. Receptacle, from which the capsules
have fallen; more *magnified.*

Tab. LXII

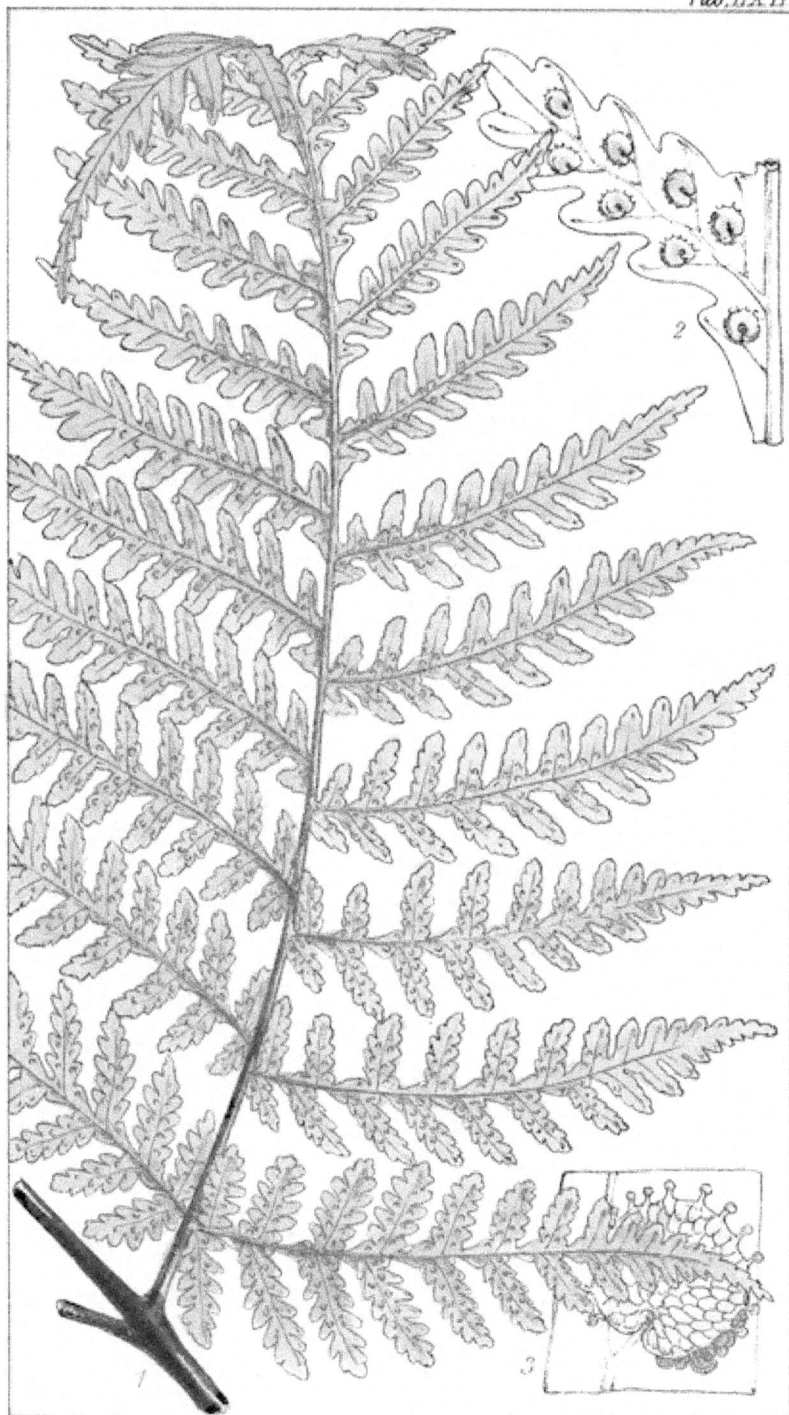

TAB. LXIII.

TRICHOMANES CELLULOSUM, *Kl.*

Caudice mediocri ascendente vel longe repente rigidissime
fibroso-radiculoso, stipitibus sparsis approximatis 2-4 uncialibus superne alatis firmis atris, frondibus 2-4 uncias longis rigidis atro-viridibus laxe cellulosis (areolis subrotundis)
tri-4-pinnatis seu potius pinnatifidis, laciniis copiosis subpatentibus linearibus seu subspathulatis obtusis integerrimis,
costa centrali crassa rigida, soris copiosis marginalibus lacinias
breves terminantibus, involucro brevi-cylindraceo marginato
ore paululum dilatato integro, collumella elongata longe ex-
serto crasso.

Trichomanes cellulosum, *Kl.* in *Linnæa*, 18, p. 531. *Kze. in
Bot. Zeit.* 5, p. 418. *J. W. Sturm, in Mart. Fl. Bras.* 23,
p. 269, t. 18, f. 13, *Van den Bosch, Syn. Hymenoph*, p. 25.

Tr. filiforme, *J. W. Sturm, in Mart.* l. c. p. 269, t. 18, f. 14?
(*The two figures in Martius are represented in "nature-
printing," excluding the cauder, and give no idea of the
fructification or of the nature of the frond*).

HAB. Tropical America; Roraima, British Guiana, *Robt.
Schomburgk,* and Kunnuku Mountains, *Rich. Schomburgk,* n.
1184; Valley of the Amazon; Baña of Rio Negro and San
Carlos, N. Brazil, *R. Spruce,* n. 1399, 2278 (n. 873, according to Sturm,) and n. 2838, segments of the fronds a little
broader and more opaque.

Few who have not studied the numerous individuals of the
family of *Hymenophyllaceæ*, can have an idea of the difficulty
that attends their correct discrimination. Happily Dr. van den
Bosch, for some time distinguished by his valuable writings
on the Mosses of the Netherlands possessions in the Malay
Archipelago, has taken up this beautiful group of Ferns, with
an amount of knowledge and of zeal which augurs well for the
Monograph, and many of the species are to be illustrated by
figures. The present species seems to be peculiar to B. Gui
ana and to the valley of the Amazon; and judging by the
references to Spruce's specimens, *T. filiforme* of J. W. Sturm
in Martius is only a var. of it.

TAB. LXIII. Exhibits a plant of Trichomanes cellulosum,
Kl.;—*nat. size. f.* 1. Portion of a pinna with fructification.
f. 2. Involucre laid open to show the capsules at the base of
the columnar receptacle;—*magnified.*

CENT. 2. T. 63.

Tab. LXVI

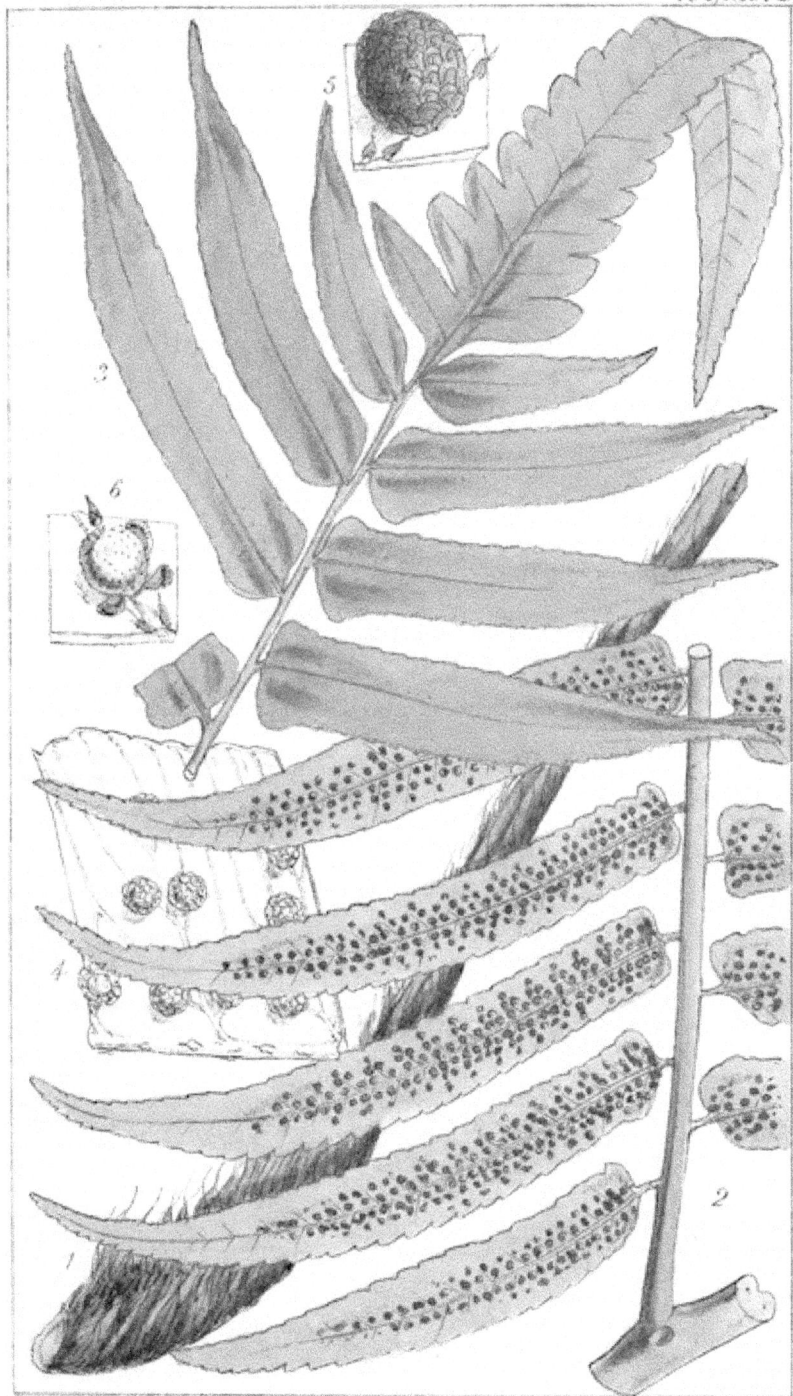

TAB. LXVII.

Nephrodium (Lastrea) Fijiense, *Hook.*

Fronde ampla bipedali et ultra submembranacea firma siccitate fusco-viridi late ovato-acuminata bipinnata, pinnis
primariis remotis ovatis acuminatis petiolatis (supremis
exceptis), secundariis seu pinnulis oblongo-linearibus acutis
sessilibus profunde fere ad costam pinnatifidis, segmentis
oblongo-ovatis ciliatis obtusis inferioribus pinnatifido-lobatis
reliquis subintegris, soris biserialibus in dorso venularum,
involucris orbiculari-cordatis pilis clavatis ciliatis, stipite
rachibusque ferrugineo-paleaceo-hirsutis.

Hab. Naviti Levu, Fiji Islands, *Milne*, n. 159, on mountains,
not common.

I am not aware of any described species of the *Lastrea*-
group to which this can be referred. There is nothing to
correspond with it among the Lastreas in Brackenridge's
Ferns of the Fijian group.

Tab. LXVII. *Fig.* 1. Primary pinna with a portion of
the main rachis, and *f.* 2. Apex of a frond of *Nephrodium
Fijiense;—natural size. f.* 3. Fertile segment of a frond,
with sori; *f.* 4. Single sorus. *f.* 5. Portion of an involucre;
all more or less *magnified*.

TAB. LXVIII.

Asplenium (Euasplenium) induratum, *Hook.*

Caudice repente radicante, stipitibus approximatis atris biuncialibus squamosis, squamis subulato-setaceis atris basi
laciniatis, frondibus 4-6 uncialibus erectis lanceolatis rigido-
chartaceis olivaceo-viridibus pinnatis, pinnis subsessilibus
horizontalibus 4-5 lineas longis 2½ latis semi-ovatis basi
superne productis apice obtusis nunc leviter falcatis crenato-serratis, venis remotiusculis, soris obliquis biserialibus
linearibus, involucris rigide coriaceis fuscis, rachi atro-villosa.

Hab. Interior of Naviti Levu, Fiji Islands. *Milne*, n. 131.

This will perhaps rank near *Aspl. hirtum* of the *Furcatum*-
group of *Euasplenium*, of which it may possibly prove to be
a small variety. All the specimens however are very uniform,
and none exceeds in size those here represented. Some of
them have lost their lower pinnæ, as if they separated at a
joint, as in so many species of *Nephrolepis*. This may be
an accidental circumstance, arising probably from a season of
unusual heat and drought; so that I have not ventured to
indicate it in our figure.

Tab. LXVIII. Plant of *Asplenium induratum*;—*natural
size. Fig.* 1. Scale from the stipes, and *f.* 2. Pinna with
sori;—*magnified. f.* 3. Sorus, more *magnified.*

TAB. LV.

HEMIONITIS LANCEOLATA, *Hook.*

Stipite 8-10-pollicari crasso badio nitido, fronde subæque
longa coriaceo-membranacea firma lato-lanceolata acuminata anguste marginata costata, costa valida subtus prominente fusca nitidissima, venis patentibus ubique anastomosantibus areolas oblongas hexagonas formantibus, venis
omnibus soriferis.

HAB. Mountain woods, interior of Naviti Levu, Fiji Islands,
Milne, in Capt. Denham's Voyage of the Herald.

This quite accords in generic character with the original
Hemionitis of Linnæus, and is the only known species of that
genus, with quite undivided fronds. Indeed without the sori,
and without observing the venation, this plant would pass
for an *Acrostichum* of the *Elaphoglossum* group. Only 2 specimens of it appear to have been preserved by Milne, and
these are destitute of caudex: but in other respects they are
very perfect. The veins are all connected just within the
margin by a longitudinal vein as represented at our fig. 2.

TAB. LV. Hemionitis lanceolata; *nat. size. Fig.* 1. Portion
of fertile frond; *magnified. f.* 2. Section showing the marginal venation, *magnified. f.* 3. Sori more highly *magnified.*

CENT. 2. T. 55.

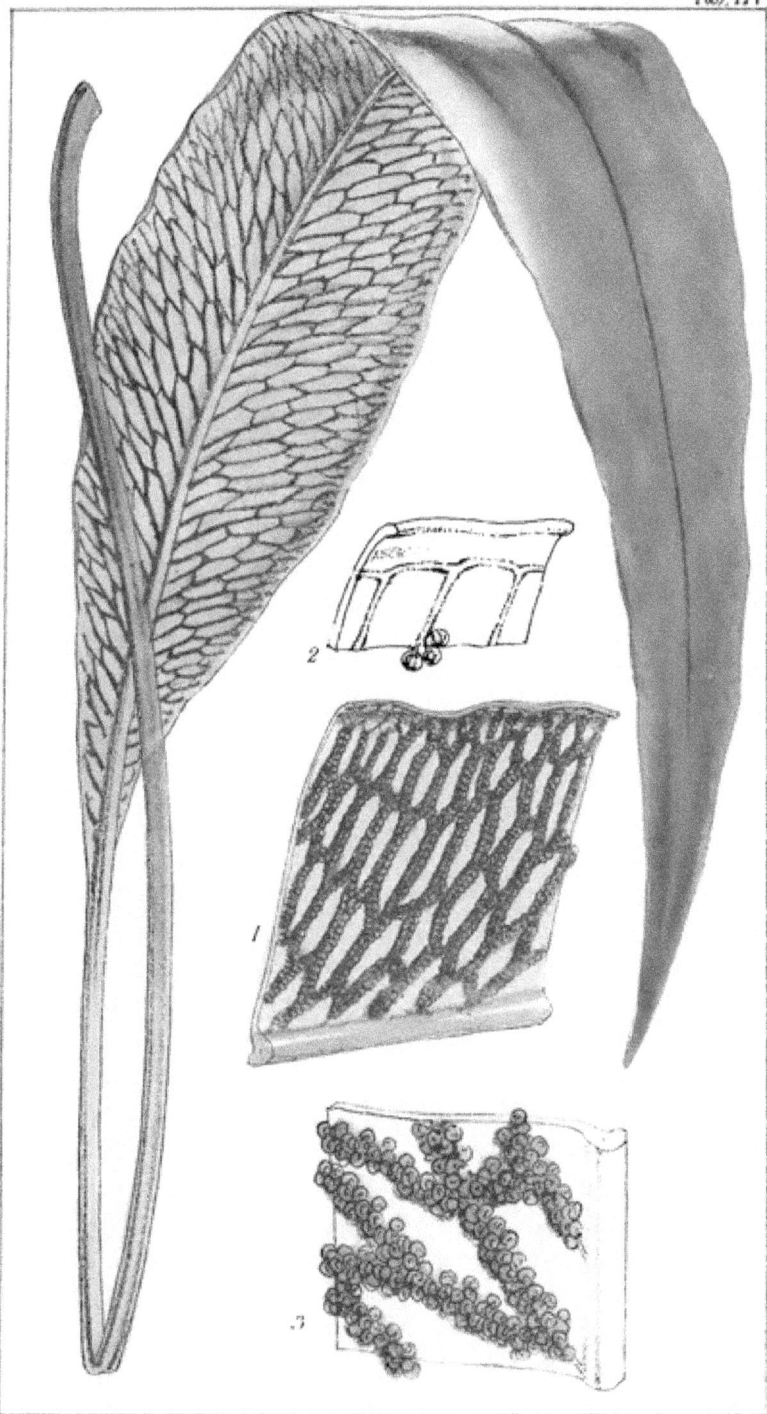

Tab. LV.

DAVALLIA (ODONTOLOMA) LA PEYROUSII, *Hook.*

Caudice brevi repente, stipitibus brevibus bipollicaribus aggregatis, frondibus 6 uncias ad sesquipedalem herbaceis lanceolatis pinnatis basi angustatis, pinnis numerosis pollicaribus horizontalibus semiovatis falcatis, parte inferiore integra, superiore profunde pinnatifida, laciniis 6-8 anguste cuneatis apice truncatis subsinuatis rarissime bi-trifidis, venis simplicibus v. furcatis, soris infra apicem sitis, involucris reniformi-oblongis transversis membranaceis subintegris, rachi stricta straminea nitida.

HAB. Island of Vaniholla or Pitt's Island, S. Pacific Ocean, the site of the disastrous shipwreck of La Peyrouse in 1788, *Mr. Charles Moore.* Naviti Levu, Fiji Islands, damp places on mountains, *Milne,* in Capt. Denham's Voyage of the Herald.

This *Davallia,* though quite distinct from, is nevertheless nearly allied to, the very elegant *Davallia Blumeana,* Hook. (Sp. Fil. p. 177, t. 54. A) from Java. That species is bipinnate; this simply pinnate. Our first specimens received were from Mr. Moore, gathered in the island which was the scene of the shipwreck of the unfortunate La Peyrouse, from which circumstance I have derived the specific name.

TAB. LVI. Davallia La Peyrousii ; *nat. size.—Fig.* 1. Sterile segments; *magnified,* and *f.* 3 & 4, fertile segments; more *magnified.*

CENT. 2. T. 56.

Tab LVI.

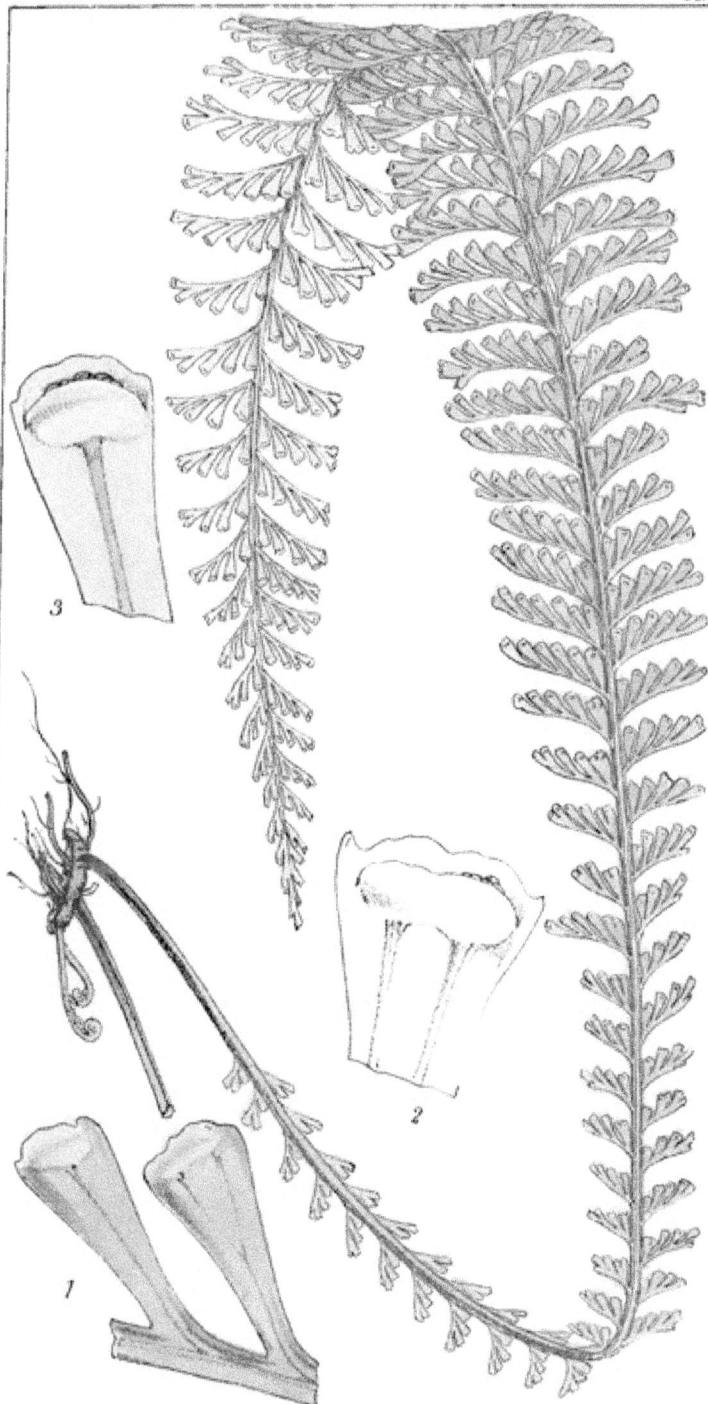

Tab. I.1

TAB. LII.

Nothochlæna ferruginea, *Hook.*

Caudice repente longe fibroso bulbillis ovatis squamosis demum frondiferis onusto, stipitibus aggregatis 2-4 pollicaribus rigidis tomentosis demum nudis aterrimis, frondibus 6-8-10-pollicaribus erectis coriaceis firmis lanceolatis pinnatis, pinnis horizontalibus sessilibus subsemiunciam longis obtusis pinnatifidis, supra villosis subtus densissime ferrugineo-v. albopannosis, laciniis utrinque 6-8 breviusculis ovato-oblongis uniformibus, marginibus magis minusve revolutis subinvolucriformibus, soris aterrimis.

Cheilanthes ferruginea, *Willd. Herb. Kaulf. En.* p. 209. *Metten. Chil.* p. 23.

Nothochlæna rufa, *Pr. Rel. Hænk.* 1. p. 19. *Liebm. Fil. Mex.* p. 62.

Nothochlæna tomentosa, *Desv. Journ, Bot.* 3. p. 92.

Nothochlæna trichomanoides, *Mart. et Gal. Fil. Mex.* p. 45, (*not Br.*)

Hab. Peru (*Vahl, Herb. and Poeppig*); Columbia (*Moritz*); Guatemala, and Mexico are given as localities for this Fern; and in the latter country it appears to be a frequent inhabitant of the mountains, at elevations varying from 3 to 5000 feet. Jamaica, St. Andrew's Parish, near Shallotenburgh Great House, elev. 3-4000, on rocks and shingly soil fully exposed, *Mr. Nathaniel Wilson.* If Willdenow's *Acrostichum Bonariense* be the same plant, as implied by Mettenius' synonyms, it is a native of Buenos-Ayres; but this requires further confirmation. May not that be the *Nothochlæna hypoleuca?* a nearly allied species, known to be a native of Chili.

This is a very elegant Fern, and now for the first time found in the West Indies by Mr. Wilson, the indefatigable superintendent of the Botanic Garden in Jamaica. Generically this borders very closely upon *Cheilanthes*, and has nearly as good a claim to rank with the one Genus as with the other.

Tab. LII. *Fig.* 1. Plant of the ferruginous state of *Nothochlæna ferruginea: f.* 2. The white state; *natural size. f.* 3. Segment with sori, *magnified; f.* 4. Portion of a fertile segment, more highly *magnified.*

Cent. 2. t. 52.

Tab. LII.

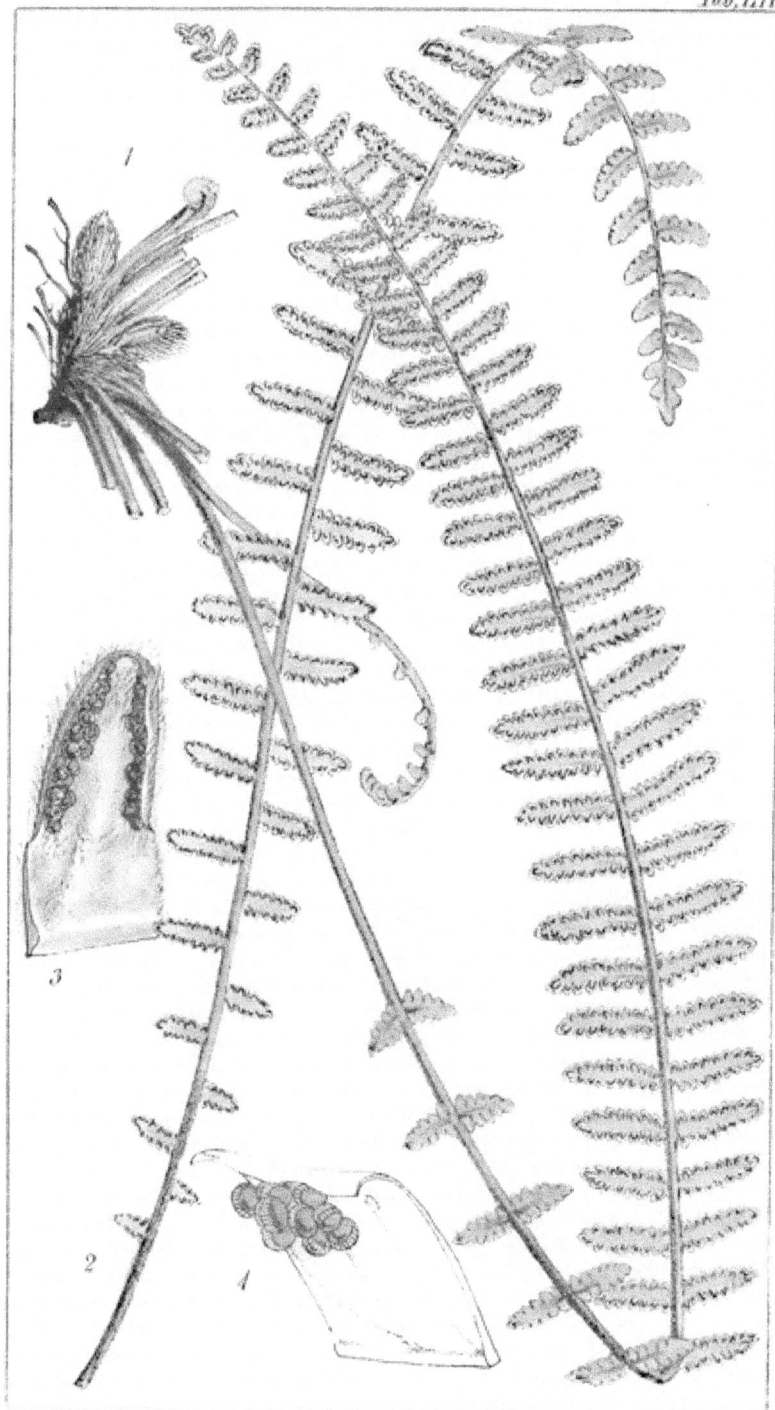

TAB. LIII.

DAVALLIA (EUDAVALLIA) MOOREI, *Hook.*

Caudice repente ferrugineo-tomentoso-squamuloso, stipitibus
sparsis firmis flexuosis semipedalibus ad pedalem basi squamulosis, frondibus coriaceis subdeltoideo-acuminatis sparsis
3-4-pinnatis, pinnis primariis ovato-lanceolatis valde acuminatis pinnatifidis ultimis lanceolatis dentato-pinnatifidis
dentibus soriferis, soris intramarginalibus cupuliformibus.

HAB. Canalla, New Caledonia, *Mr. Charles Moore*, n. 5.

This species of *Davallia*, belonging to the group or section
Eudavallia, possesses no very strikingly marked characters,
and yet cannot be referred to any known species. It comes
too from a country in the Australian S. Pacific Ocean,
which would doubtless yield a good harvest of new Ferns,
could it be fully investigated by the Botanist.

TAB. LIII. Plant of *Davallia Moorei*; natural size. *Fig.* 1.
Ultimate pinna, fertile. *f.* 2. Single sorus;—*magnified.*

TAB. LIV.

DAVALLIA (§ DAREOIDEÆ) FŒNICULACEA, *Hook.*

Frondibus sesquipedalibus (et ultra?) subcoriaceo-herbaceis latè ovatis acuminatis 4-pinnatis, divisionibus primariis circumscriptione ovato-lanceolatis tenui-acuminatis 6-8 pollicaribus, secundariis tertiariisque lanceolatis segmentis angustissime linearibus ultimis vix lineam longis subclavatis, soris copiosis, involucris solitariis suburceolatis ad marginem interiorem infra apicem oblique sitis.

HAB. Naviti Levu of the Fiji Islands, in woods on mountains, *Milne*, in the Voy. of the Herald, under Capt. Denham, R.N.

The most finely and deeply cut of all the genus *Davallia*. The resemblance to fennel leaves is indeed weakened by the very copious sori, which are situated on the interior margin of most of the ultimate segments, and so prominent, and so oblique in direction, that a fertile segment not inaptly resembles a tobacco-pipe, with a crest or wing on one side. A very obscure costa passes through all the segments ; indeed the ramification is of that character that it may be as correctly described 3-4-pinnatifid as pinnate.

TAB. LIV. *Fig.* 1. Primary branch or division of Davallia fœniculacea ; *natural size. f.* 2. Ultimate segments, one fertile, —*magnified; f.* 3. Fertile segment, more *magnified.*

CENT. 2. T. 54.

Tab. LIV.

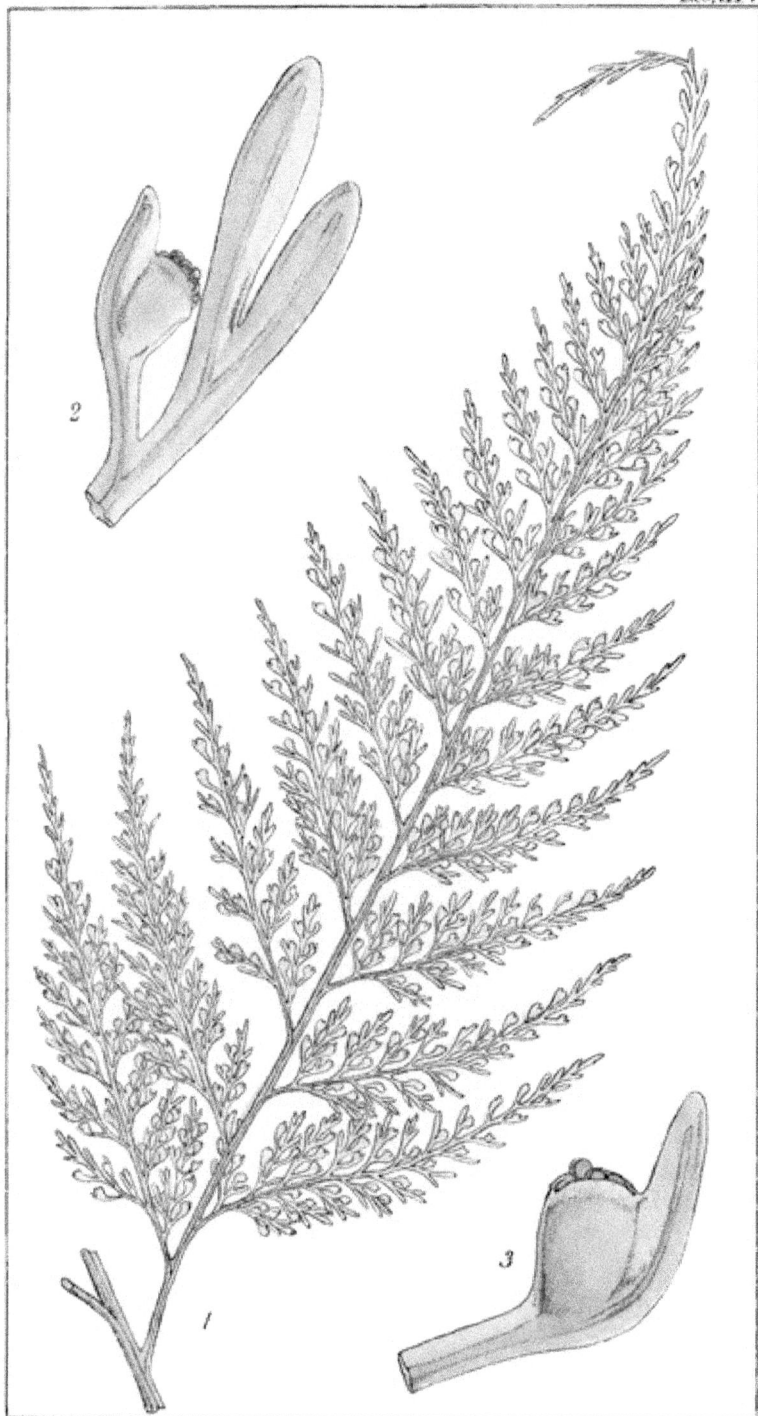

TAB. XLI.

ASPLENIUM (EUASPLENIUM) MONTEVERDENSE, *Hook.*

Caudice parvo fibroso-radicante, stipitibus cæspitosis 1-2 uncias longis lurido-castaneis, frondibus 4-6 uncias longis lato-lanceolatis acuminatis inferne attenuatis membranaceis pallide viridibus demum subcoriaceis atro-viridibus tripin-natis, pinnis omnibus subpetiolatis, primariis patentibus 1-1½ unciam longis oblongo-ovatis obtusis, secundariis lato-cu-neatis (infimis brevibus pauci-pinnatis) plerisque bi-trifidis segmentis brevibus acutis modice subincurvis, venis costæ-formibus apice clavatis, soris in lobos solitariis vix dareoideis, involucris parvis lineari-oblongis flavo-virescentibus valde membranaceis, rachibus viridibus compressis marginatis.

Asplenium Monteverdense, *Hook. Sp. Fil. vol.* 3. *p.* 195.

HAB. Under overhanging rocks near Mount Verde, on the eastern side of Cuba, *C. Wight. n.* 1029.

Of this species the affinity is perhaps with *Aspl. varians,* Hook. & Grev., but the very short stipites, the fronds attenu-ated at the base, the different form of the pinnæ and their more distinct petioles will readily distinguish it; or, more nearly, with some of the many forms of *Aspl. cicutarium.* In the old state of fructification the plant might be taken for a *Gymnogramme,* for the copious capsules soon cover and conceal the involucres.

Fertile fronds ; *nat. size. Fig.* 1. Pinnule with sori. *f.* 3. Single sorus : *magnified.*

Tab. XLI.

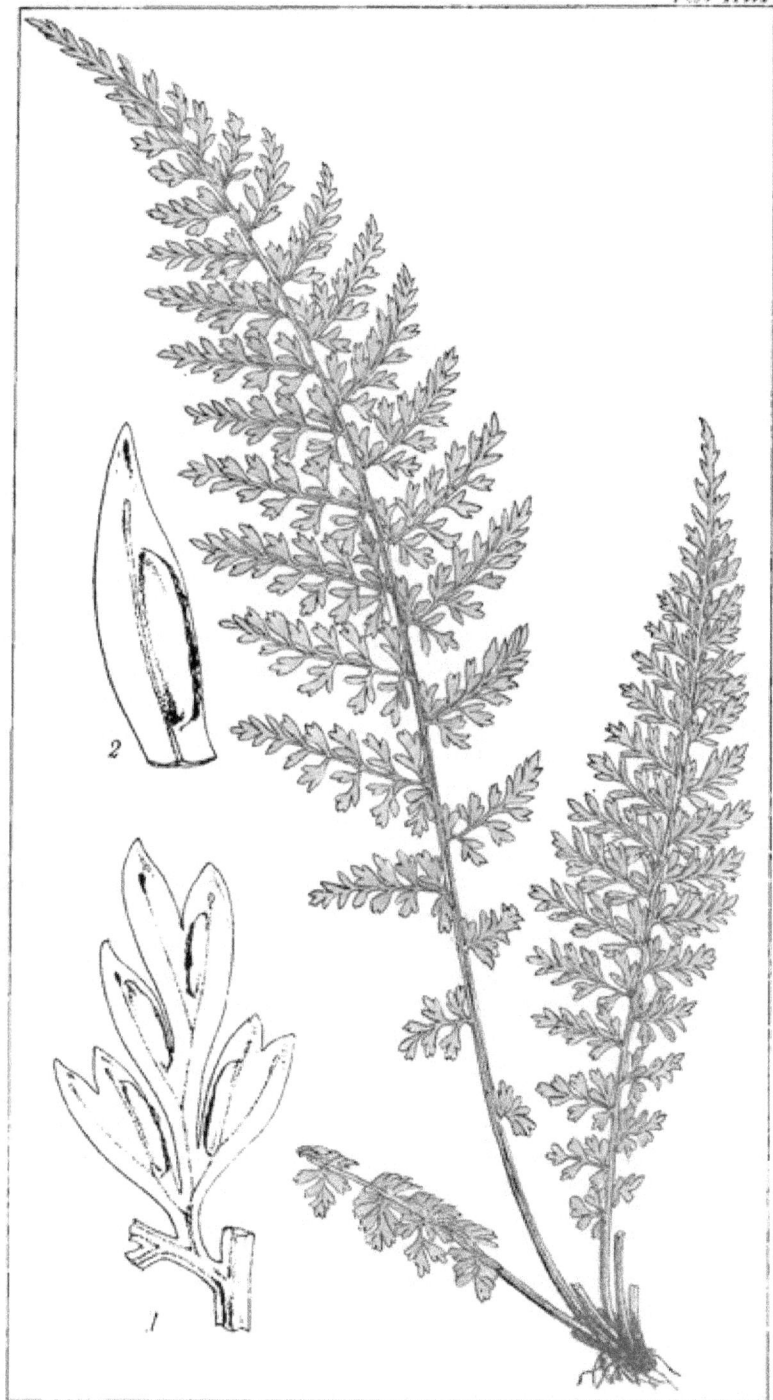

TAB. XLII.

Asplenium (Euasplenium) prolongatum, *Hook.*

Caudice parvo ascendente radicante vix squamoso, stipitibus cæspitosis 2-4 uncias longis stramineis subcompressis, frondibus 4-5-uncialibus ad pedalem coriaceis seu subchartaceis oblongis vel lineari-oblongis sæpe falcatis bi-subtripinnatis, rachi apice prolongata caudiformi nuda radicante 1-2 uncias longa, pinnis primariis 1-1½ uncias longis horizontaliter patentibus sæpe approximatis semiovatis obtusis petiolatis semipinnatis (seu pinnulis hinc longioribus numerosioribus), pinnulis 3-4 lineas longis plerisque simplicibus integris raro furcatis ad basin superiorem bi-tripartitis, venis costiformibus, soris oblongis submarginalibus, involucris firmis membranaceis colore frondis.

Asplenium prolongatum, *Hook. Sp. Fil.* 3. *p.* 209.

HAB. On Trees, East Indies, Mishmee, *Griffith, Simons, n.* 235; Khasya, *Hook. Fil. et Thomson;* Bhotan, *Booth;* Ceylon, *Mrs. Genl. Walker, Gardner, n.* 1348; Tsus Sima, Strait of Korea, *Wilford.*

A very elegant and well marked species, retaining its characteristic distinctions in all the specimens from the several localities above mentioned. Nearly every one exhibits the remarkable prolongation of the rachis, rooting at the apex and often very proliferous there. Of the primary pinnæ the longest and most numerous pinnules are always on the upper half.

Fertile plant; *nat. size. Fig.* 1. Pinna. *f.* 2. Sorus: *magnified.*

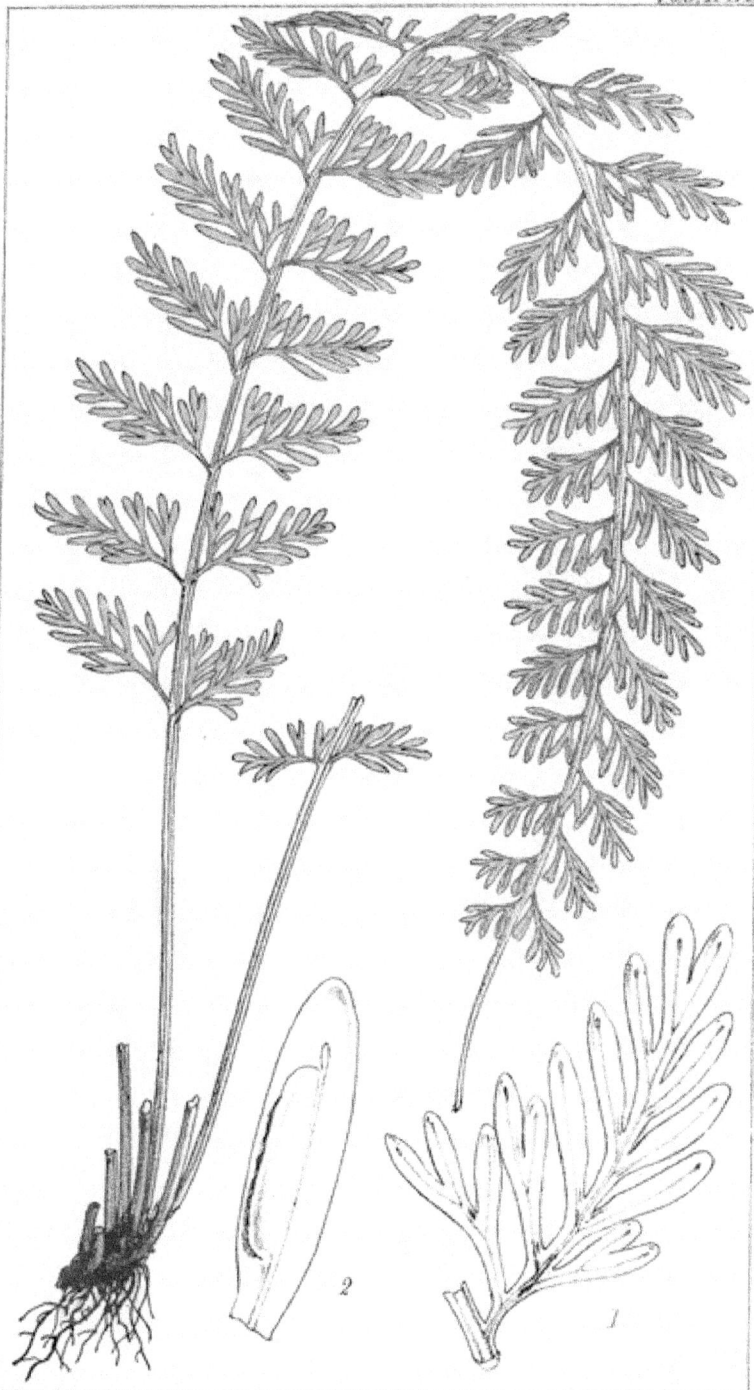

Tab. XLII

TAB. XLIII.

Asplenium (Athyrium) medium, *Carm.*

Caudice "6-7 uncias longo," stipitibus 5-6 unciam longis stramineis basin versus incrassatis paleis longissimis angustissimis ferrugineis flexuosis obsitis, frondibus spithamæis rigide subcoriaceis deltoideis acutis bipinnatis, pinnis sæpe oppositis horizontalibus approximatis sessilibus 3 uncias longis ovato-lanceolatis, pinnulis ½-¾ unciam longis lato-lanceolatis sessilibus pinnatifidis acutis vix auriculatis segmentis ovato-oblongis serratis, venis pinnatis dichotomis, soris copiosis, involucris membranaceis reniformibus sinuatis margine erosis, rachi universali crinita.

Asplenium medium, *Hook. Sp. Fil.* 3. *p.* 228.

Aspidium medium, *Carm. in. Linn. Trans.* 12. *p.* 311.

Aspidium intermedium, *Carm. Mss. in Herb. Hook.* Athyrium, *Moore. Ind. Fil. p.* 96.

Hab. Tristan d'Acunha, on the Table-land; *Dr. Carmichael, in Herb. nostr.*

A very distinct species, peculiar, as far as yet known to the island just mentioned. Its discoverer has noted, "stem (caudex) about 6 inches high, crowned with a circle of fronds from 9 to 12 inches high."

Fertile plant; *nat. size. Fig.* 1. Pinnule with sori. *f.* 2. Single sorus. *f.* 3. Setiform scale from the stipes: *magnified.*

Tab. XLIII.

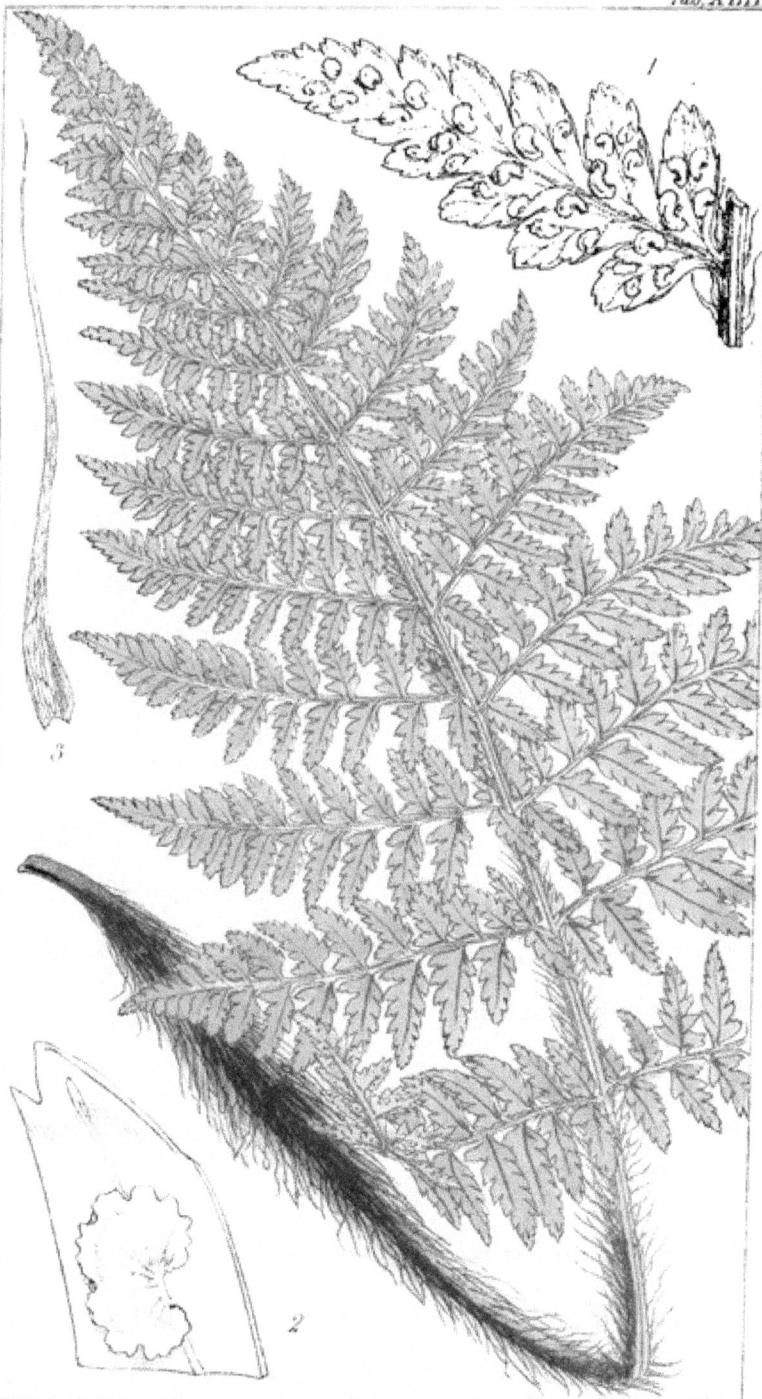

TAB. XLIV.

ASPLENIUM (ATHYRIUM) NIGRITIANUM, *Hook.*

Tota planta siccitate nigra, caudice brevi robusto erecto
squamis copiosis atro-ferrugineis subulatis paleaceo, stipitibus cæspitosis robustis spithamæis rachique dense fusco-
villosis, frondibus 1-1½ pedem longis rigide coriaceis oblongo-lanceolatis acuminatis basi attenuatis bi-tripinnatis,
pinnis infimis remotis, reliquis magis approximatis horizontalibus 2½ vix 3 uncias longis lanceolatis acuminatis,
pinnulis 2-3-lineas longis omnibus petiolulatis oblique
rhomboideis obscure et obtuse auriculatis seu inæqualiter
bilobis serratis, pinnulis infimis ternatis seu subpinnatis
rachin imbricantibus, venis obscuris subflabellatim dichotomis, soris 1-5 in singula pinnula parvis lato-oblongis, involucris convexis integerrimis fere nigris.

Asplenium Nigritianum, *Hook. Sp. Fil.* 3. *p.* 223.

HAB. Prince's Island, Fernando Po, *Barter in Baikie's 2nd
Niger Exped. n.* 1898.

A distinct and very peculiar *Asplenium*, which has no near
relationship to any species known to me; remarkable for its
firm, rigid texture, very stout stipes and main rachis, shaggy
with woolly, hair-like scales, and for the very black colour of
the whole plant when dry. I place it with some hesitation
among *Athyria*, on account of a certain peculiarity of habit,
and of the convex involucres, which however are very firm
and coriaceous, and nearly of the same colour as the (dried)
frond.

Portion of the base and apex of a fertile plant; *nat. size.
Fig.* 1. Inferior pinnule from a sterile pinna. *f.* 2. Fertile
pinna with sori: *magnified.*

CXM. 2. t. H.

Tab. XLIV

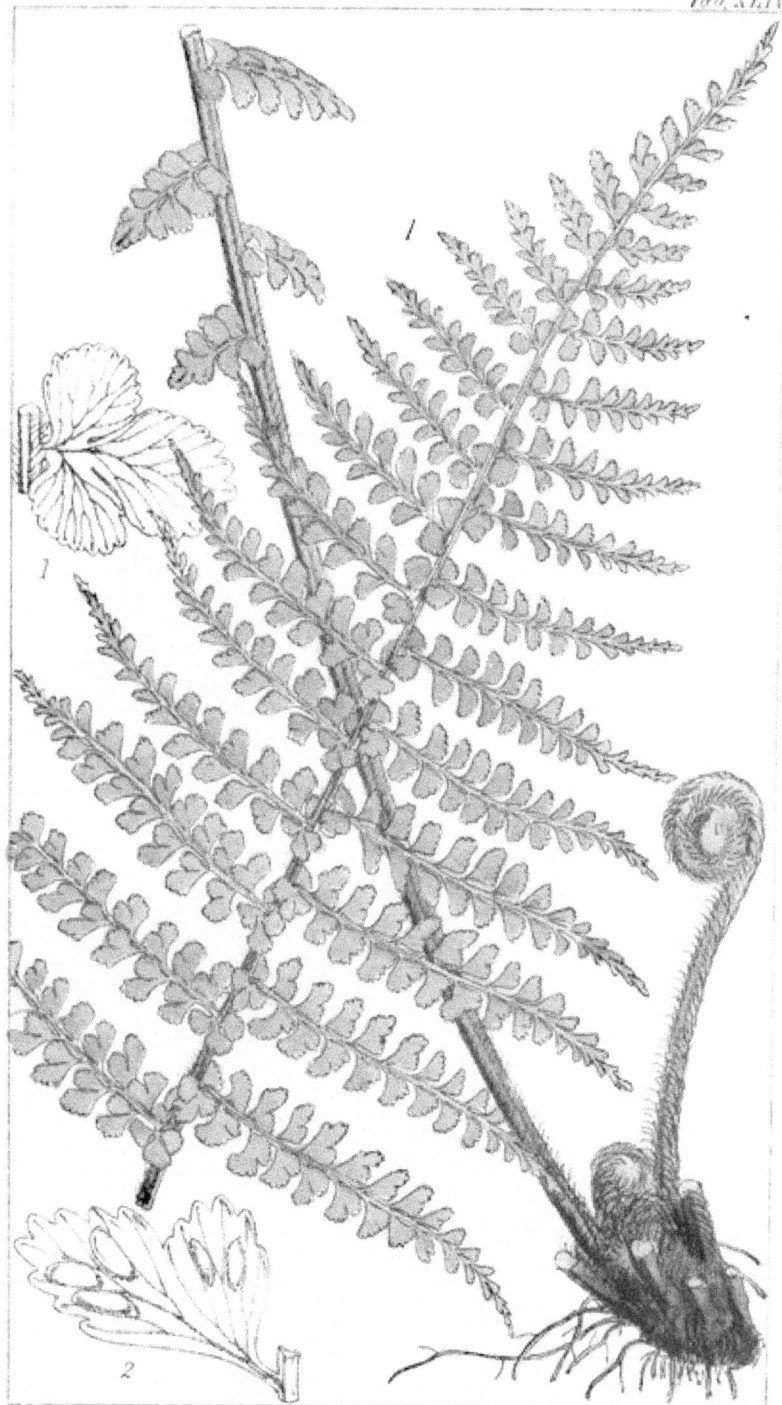

TAB. XLV.

ASPLENIUM (EUDIPLAZIUM) THWAITESII, *A. Braun.*

Caudice longo repente atro radicante, stipitibus sparsis 4 uncias ad spithamæam longis rachique pilis crispatis squamisque lanceolatis membranaceis vestitis, frondibus spithamæis ad pedalem ovato-lanceolatis acuminatis membranaceis pinnatis apice pinnatifidis, pinnis 1½-3 uncias longis approximatis sessilibus horizontalibus obtusis rectis profunde fere ad rachin pinnatifidis, lobis brevi-oblongis obtusis apice dentatis, venis pinnatis simplicibus v. furcatis, soris singulo lobo in seriebus duabus parvis lineari-oblongis, involucris pallide fuscis membranaceis convexis suberosis nunc diplazioideis, costis et venis supra subglandulososparseque pilosis.

Asplenium Thwaitesii, *A. Braun, Ind. Hort. Berol.* 1857 *Metten. Asplen. p.* 183. *Hook. Sp. Fil.* 3. *p.* 250.

HAB. Ceylon, *Gardner, n.* 1343, *Thwaites.*

A remarkable and well defined species in its long, creeping, subterranean, blackened caudex, and the densely tomentose and paleaceous stipites and main rachises.

Fertile frond; *nat. size. Fig.* 1. 2. 3. Portions of pinnæ with sori: *magnified.*

CASL. Z. T. F.

Tab. XLIV.

TAB. XLVI.

ASPLENIUM (EUDIPLAZIUM) VESTITUM, *Hook.*

Caudice? stipitibus robustis paleaceis squamosis, squamis infimis maximis ovatis acuminatis atro-fuscis nitidis denticu-
latis, frondibus amplis submembranaceis fusco-viridibus
sesquipedalibus ovato-lanceolatis pinnato-pinnatifidis apice
pinnatifidis, v. bi-tripedalibus latissime ovatis bipinnatis,
pinnis omnibus patentissimis oblongis petiolatis, primariis
distantibus semi-pedalibus ad spithamæam acutis apice
pinnatifidis, pinnulis elliptico-oblongis 2 uncias longis fere
unciam latis sæpe obtusissimis basi truncatis lobato-pinnatifidis superioribus subintegris serratisque, lobis obtusis vel
subangulatis, venis pinnato-fasciculatis, soris linearibus
copiosis infimis præcipue diplazioideis, rachibus villoso-
squamulosis.

Asplenium vestitum. *Hook. Sp. Fil.* 3. *p.* 263.

Diplazium vestitum. *Pr. Epimel. Bot. p.* 87.

D. extensum. *J. Sm. in Hook. Bot. Journ.* 3. *p.* 407, *in part,*
(*name only.*)

HAB. Isle of Samar, Phillippines, *Cuming. n.* 336.

Our younger specimens, as they appear to be (yet bearing
copious sori), are simply pinnated, the larger and older ones
bipinnate, with large very distant primary pinnæ.

Fig. 1. Stipes. *f.* 2. Portion of a fertile frond; *nat. size.*
f. 3. Portion of a fertile pinna, with sori. *f.* 4. Scale from
the base of the stipes: *magnified.*

Tab. XLVI

TAB. XLVII.

DAVALLIA (SACCOLOMA?) DENHAMI, *Hook.*

Caudice repente subulato-paleaceo, stipitibus subsparsis 3-6 uncias longis castaneis nitidis, frondibus 6-8 uncias longis subchartaceis ovatis acuminatis bipinnatis, pinnis petiolatis 2-3 uncias longis lanceolatis acuminatis remotis, pinnulis semiunciam ad unciam longis lineari-lanceolatis obtusis pinnatifidis basi sessilibus subdecurrentibus, laciniis brevissimis 1-bidentatis monosoris, involucro pyriformi parte superiore libera rotundata.

HAB. Naviti Levu, Feejee Islands, *Milne, in Voy. of Capt. Denham, n.* 116.

A very pretty new *Davallia*, which, like the following one (*D. rhomboidea*, TAB. XLVIII), I find difficult to refer satisfactorily to its group in the genus; so much do those groups gradually pass, by almost insensible characters, into others.

Fertile plant; *nat. size. Fig.* 1. Small pinnule with sori. *f.* 2. Sorus. *f.* 3. Sorus with the anterior portion of the involucre removed: *magnified.*

Tab. XLVII.

TAB. XLVIII.

DAVALLIA (CUNEATÆ) RHOMBOIDEA, *Hook. (not Wall.)*

Caudice gracili elongato repente atro nitidissimo fragili, stipitibus sparsis 3-4 uncias longis stramineis nitidis basi ebeneis, frondibus 6-8 uncias longis oblongo-lanceolatis acuminatis tenui-membranaceis pallide viridibus pinnatis, pinnis patentibus inferioribus remotis subunciam longis rhombeo-subtriangularibus longe petiolatis profunde pinnatifidis subpinnatisque lobis pinnulisve oblique obovatis inequaliter lobatis, sterilibus serrulatis, venis subflabellatim dichotomis apice soriferis, involucris orbiculari-cuneatis membranaceis apice erosa solummodo libera lobulis frondium marginis conformibus, rachibus gracilibus subflexuosis, pinnis supremis sublanceolatis.

HAB. Hakodadi, Japan, *Wilford, n.* 1037.

A very elegant and peculiar species of *Davallia*, which I am disposed to refer to the section *Cuneatæ*; it is quite different from any described one, and only known to us through our collector, *Mr. Wilford.*

Fertile plant; *nat. size. Fig.* 1. Portion of a pinna with sori; *magnified. f.* 2. Sori more magnified, one representing the involucre removed: *more magnified.*

Tab. XLVIII.

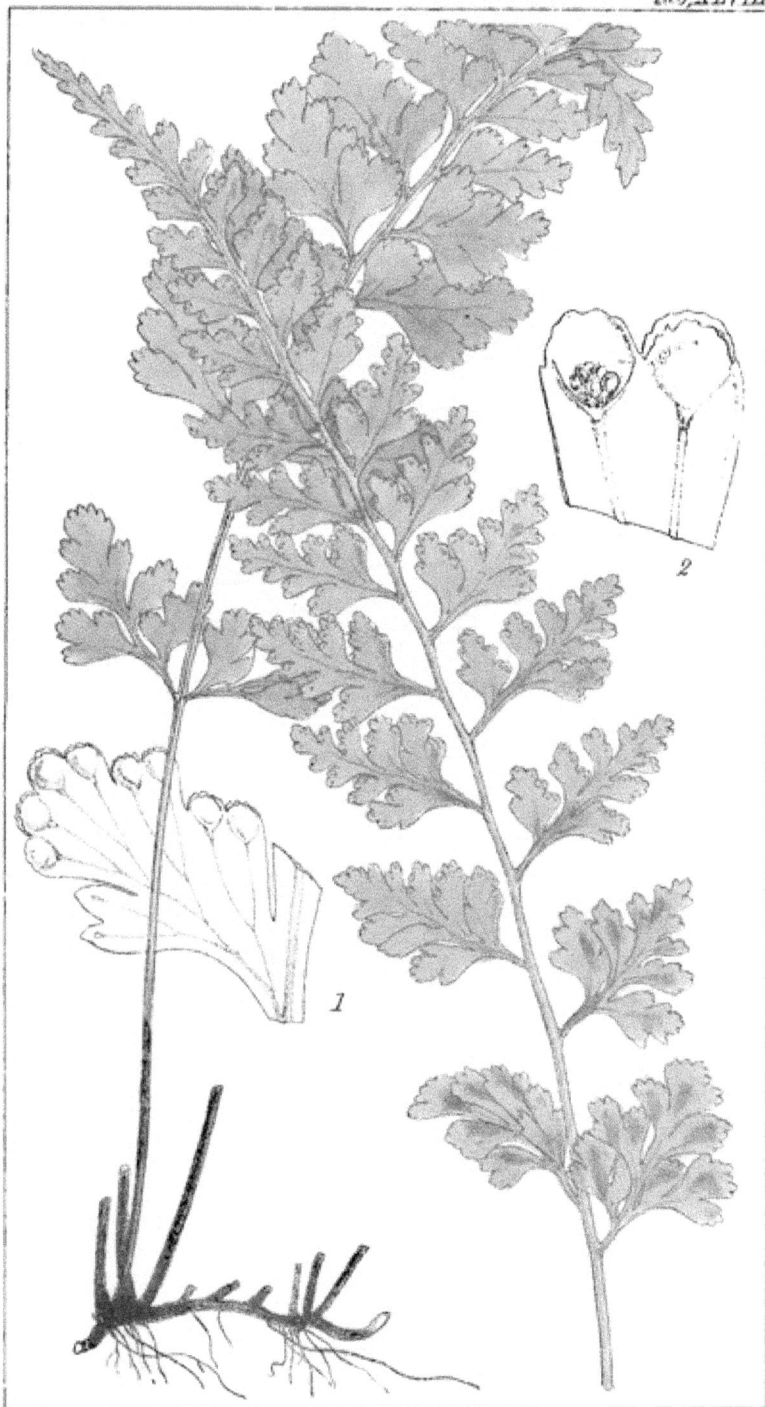

TAB. XLIX.

POLYPODIUM (PHEGOPTERIS) DECURSIVO-PINNATUM,
Van Hall.

Caudice subrepente stipitibusque semipedalibus stramineis squamis subulatis ciliatis ferrugineis paleaceis, frondibus villosulis pedalibus et ultra lanceolatis tenui-acuminatis inferne augustatis pinnatis apice pinnatifidis, pinnis horizontaliter patentibus oblongo-lanceolatis pinnatifidis, infimis brevibus liberis, reliquis lobo intermedio semicirculari coadunatis, venis pinnatis apice clavatis supra medium soriferis, soris perpaucis solitariis parvis pilis fasciculatis capsulis duplo longioribus (vix basi in membranam seu involucrum unitis) intermixtis, rachibus stramineis nitidis patentiferrugineo-villosis.

Polypodium decursivo-pinnatum, *"Van Hall in N. Verhandl. t.* 1. *Klass. v. d. Neederl. Instit. t.* 5."

Phegopteris decursivo-pinnata, *Fée gen. p.* 242. *t.* 20. A. 1. *(fragments only.)*

Aspidium decursivo-pinnatum, *Kze. Bot. Zeit.* 6. 555. *Metten. Aspid. p.* 75.

Lastrea decurrens, *J. Sm. Bot. Mag. v.* 72. *Comp. p.* 33.

HAB. Japan, *Goring* ; Port Chusan, Korea, *Wilford, n.* 920.

A very elegant and very distinct species allied to the European *Polypodium Phegopteris,* Linn. and to the *Pol. hexagonopterum,* Sw. but very different in form and in the presence of the copious, spreading, ferrugineous, subulate scales which clothe the stipites and rachis. Kunze and others consider that the hairs of the sori arise from an almost obsolete involucre, and hence they refer the plant to *Aspidiaceæ.*

TAB. XLIX. Fertile plant ; *nat. size. Fig.* 1. Portion of a fertile pinna. *f.* 2. Sorus. *f.* 3. Scale from the base of the stipes : *magnified.*

Tab. XLIX

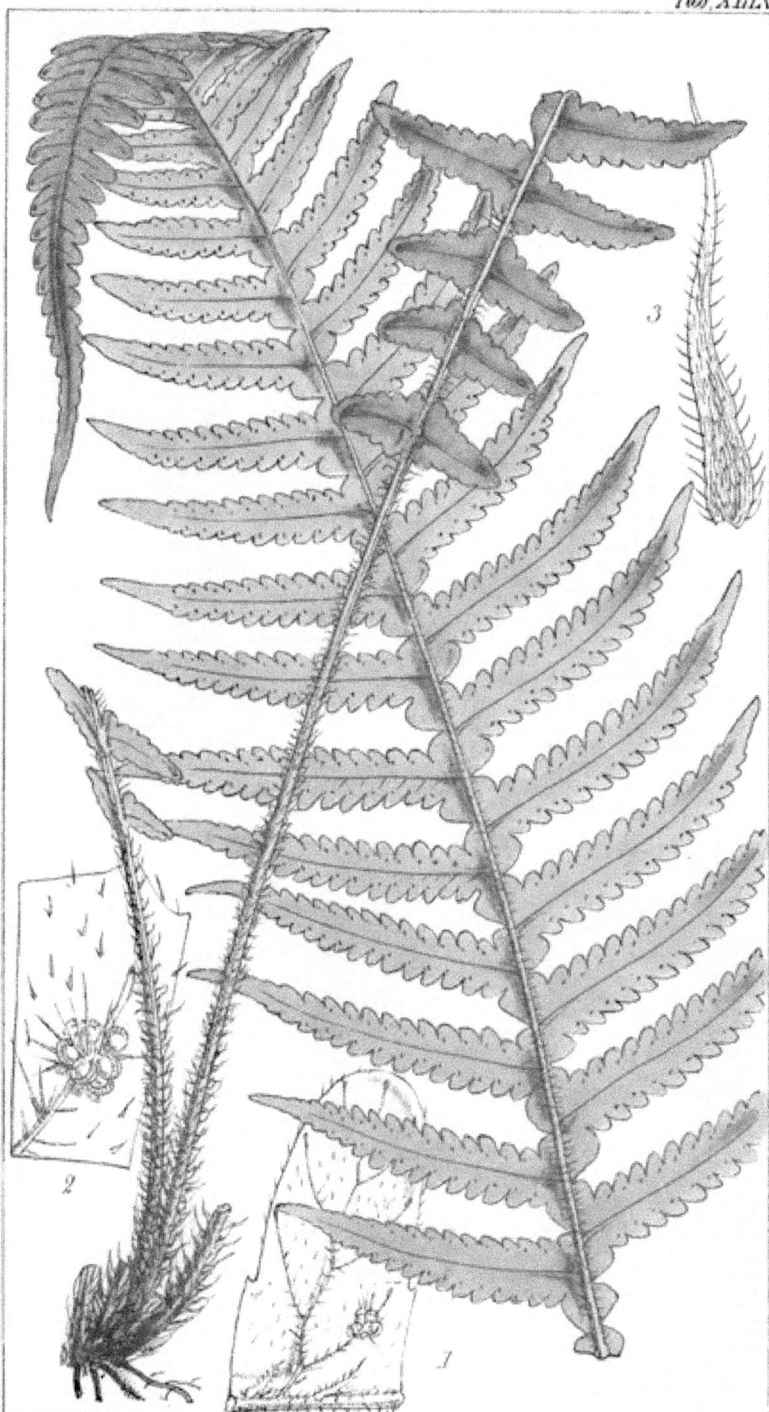

TAB. L.

ADIANTUM MONOCHLAMYS, *Eaton.*

Caudice horizontali fusco-tomentoso, stipitibus spithamæis rachibusque flexuosis castaneis nitidis, frondibus ovatis acuminatis chartaceis pallide viridibus tripinnatis, pinnulis omnibus sublonge petiolulatis obcordato-cuneatis apice crenato-serratis in sinu profundo monosoris, venis flabellatis dichotome divisis, involucro suborbiculari coriaceo atro-fusco.

Adiantum monochlamys, *Eaton, in Proceedings of Am. Acad. Arts & Sc. for* 1809. *p.* 110.

HAB Hill-sides, near Simoda, Japan, *C. Wright;* Tsus Sima, Strait of Korea, *Wilford, n.* 837.

A very elegant and unquestionably a very distinct species of *Adiantum*, well named *monochlamys* by its first describer. Among my specimens there is no instance of more than one sorus on each pinnule, and that arises from a deep sinus at the almost truncated apex.

TAB. L. Fertile plant; *nat. size. Fig.* 1. Back view of a fertile pinnule. *f.* 2 Front view showing the sorus: *magnified.*

Tab. L.

TAB. LI.

ASPLENIUM (§ ANISOGONIUM) TERNATUM, *Liebm.*

Fronde coriacea glabra ternata 5-10-poll. longa, 2½-3½ poll.
lata, stipite 2-7-pollicari, pinna media lateralibus longiore
3-5 poll. longa 1-1½ poll. lata elliptica utrinque attenuata
apice longe acuminata petiolata, petiolo 4-6 poll. longo,
margine imprimis apicem versus remote et grosse dentato ;
pinnis lateralibus oppositis inequalibus, 1-4 poll. longis
1-1½ poll. latis falcato-ellipticis brevipetiolatis acuminatis
grosse dentatis ; pagina anteriori obscure viridi posteriori
glauco-viridi ; costa media antice canaliculata, postice con-
vexa, venis utrinque prominulis nigris pluries furcatis hic
illic in areolam ellipticam anastomosantibus ; soris 3-4
lineas longis simplicibus vel diplazioideis, indusiis integris
membranaceis fu cis, stipite antice et lateribus sulcato,
postice convexo.—Rhizoma subterraneum obliquum breve
pennam anserinam crassum radiculis simplicibus validis et
intricatis et fragmentis stipitum emortuorum tectum.—
Liebm.

Asplenium, (§ Anisogonium) ternatum, *Hook. Sp. Fil.* 3. p.
265. Diplazium ternatum, *Liebm. Fil. Mex.* p. 100. *Metten.*
Aspl. p. 162.

HAB. Mexico, Distr. of Oajaca elev. 4-5000 feet, *Liebmann.*

My specimens of this plant, from the author, will be better
understood by the accompanying figures taken from them
than by words. The species has the characters in part of
Asplenium, of *Diplazium,* and *Anisogonium ;* and to those who
maintain those genera respectively, it would be difficult to
say to which of the three it has the strongest claim.

TAB. LI. Plants; *nat. size.* *Fig.* 1. Portion of a pinna
showing the venation and sori ;—*magnified.* *f.* 2. Portion of
a sorus more highly *magnified.*

Tab LXVIII

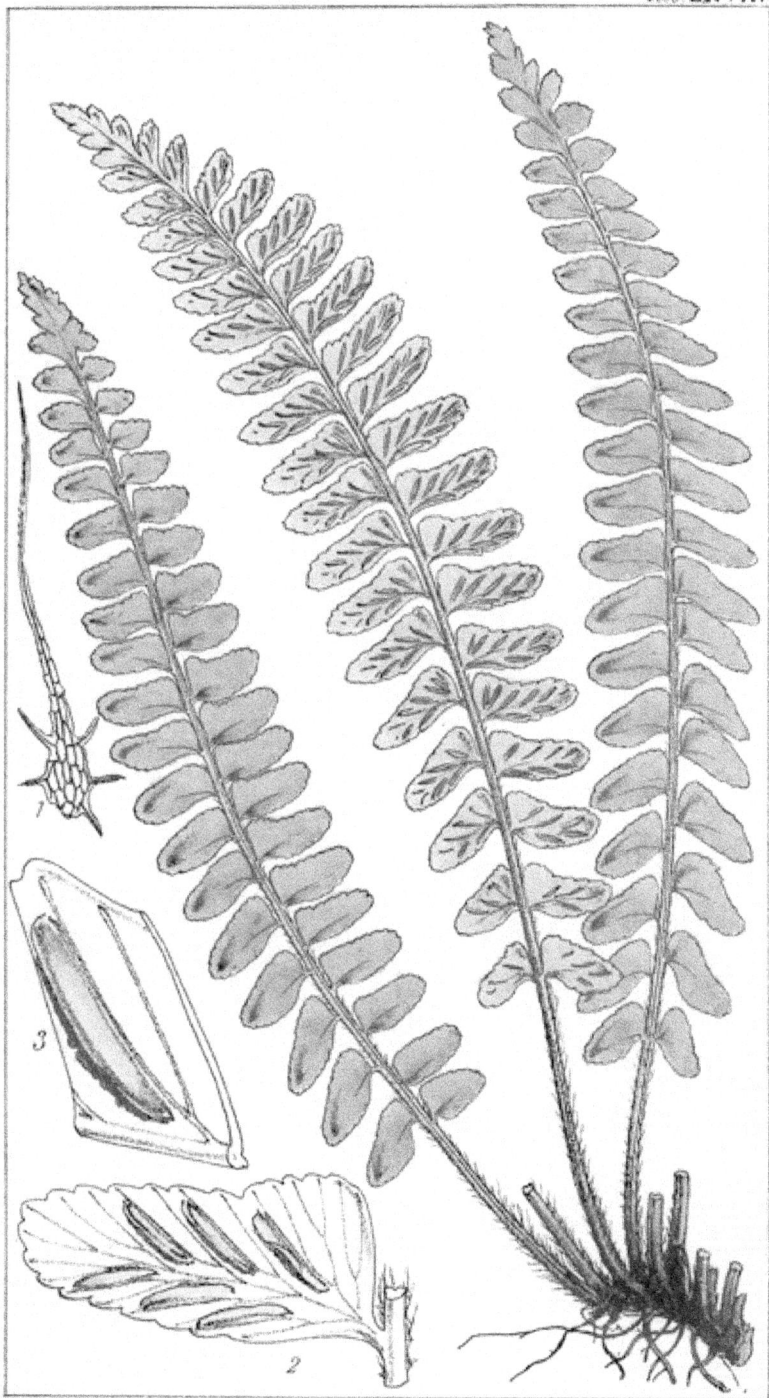

TAB. LXIX.

ASPLENIUM (EUASPLENIUM) LONGICAUDA, *Hook.*

Caudice brevi repente copiose fibroso, stipitibus spithamæis et
ultra nitidis, frondibus pedalibus ad bipedalem pergamen-
taceis (siccitate olivaceis) firmis, pinnis 5-9 late oblongo-
lanceolatis, 6-8 uncialibus acuminatis caudatis proliferis v.
cauda delapsa truncato-emarginatis margine integerrimis
v. sinuato-lobatis terminali sæpe longissima caudato-acumi-
nata et apice prolifera, costa subtus prominente, venis
remotis obliquis simplicibus v. furcatis, soris linearibus
remotis margine approximatis.

Asplenium emarginatum, *Hook. Sp. Fil.* 3, p. 100 (in part),
not Beauv.

HAB. Western tropical Africa, S. of the line, *Dr. Curror;*
Prince's Island, *Barter* in Baikie's Niger Expedition, *n.*
1900; Fernando Po, on trees, Peak Mountain, at an elevation above the sea of 3000 feet, *Gustav Mann, n.* 341.

A good suite of specimens which I now possess of this
Asplenium from the late Mr. Barter, and from Mr. Gustav
Mann, has convinced me that I have erred in uniting Dr.
Curror's plant with the *A. emarginatum* of Palisot de Beauvois: and this will be better understood when I shall shortly
give, in the present work, a figure and more perfect character of the true *emarginatum.* The two plants are certainly
nearly allied: but the present may be known by the following characters. It is a larger and less delicate plant, of a
very different and much firmer texture, resembling that of
parchment: its colour when dry is a dirty olivaceous brown.
The pinnæ are entire (not serrated) and in its normal state
gradually accuminated at the apex, and the terminal pinna
is not, though larger than the lateral ones, materially altered
in shape: but it often happens that the pinnæ are proliferous,
then the *lateral* ones are narrowly caudate at the apex and
a scaly bud forms: when this becomes a plant and falls away
a deep and wide notch takes its place. If the *terminal*
pinna is proliferous it is remarkably and gradually attenuated
(to the length of 1 or 1½ foot) and the apex copiously proliferous. The sori are always distant and are situated nearer
the margin than the costa: the reverse is the case in *A.
emarginatum.*

TAB. LXIX. Represents a proliferous frond of *Asplenium
longicauda;—natural size.* *Fig.* 1. Portion of a fertile pinna,
with a sorus, *magnified.*

CEST. 2. 1. 69.

TAB. LXX.

ANTROPHYUM GALEOTTII, *Fée.*

Caudice vix repente subnullo, radicibus fibrosis copiosis et
  cæspitosis dense ferrugineo-tomentosis, frondibus aggregatis
  spithamæis ad pedalibus lorato-lanceolatis acuminatis flaccidis sessilibus, costa latiuscula obscura, venis immersis in-
  distinctis anastomosantibus areolis oblongis, soris linearibus
  submarginalibus simplicibus vel elongatis ramosis nunc
  anastomosantibus.

Antrophyum Galeottii, *Fée, Antroph.* p. 51, t. 5, f. 4 (1832).

Antrophyum falcatum, *Mart. et Gal.* p. 49, t. 12 *(not Blume).*

Antrophyum ensiforme, *Hook. in Benth. Plantæ Hartweg.*
  p. 73 (1839).

Scoliosorus ensiformis, *Moore, Ind. Fil.* p. xxix.

Hab. Mexico, *Galeotti, Hartweg,* n. 522. Guatemala, *Skinner,*
  and on mountains of Vera Paz, elev. 3500-5000 feet.
  Osbert Salvin, Esq.

This is probably a very rare Fern, apparently peculiar to
Mexico, where it has been detected by Galeotti and Hartweg,
and Guatemala, whence we have specimens from Mr. Skinner,
and most beautiful ones from Mr. Salvin. Like some other
Antrophya it has perfectly sessile fronds, and since we have
satisfied ourselves that it has anastomosing venation, and
sometimes anastomosing sori, though that is by no means
inconsistent with *Antrophyum,* we retain it in the Genus of
which it is a true member. Mr. Moore (under his Genus
Scoliosorus) says, "This plant having neither netted veins
nor netted sori, cannot belong to *Antrophyum,* and is quite
different from every other established Genus.

Tab. LXX. Fertile plant of *Antrophyum Galeottii,* Hook.
Fig. 1. Portion of a fertile frond, with sori;—*magnified.*

Tab. LXX.

TAB. LXXI.

GRAMMITIS (LOXOGRAMME) SALVINII, *Hook.*

Caudice repente squamoso radicibusque fusco-tomentosis, frondibus remotis subspithamæis submembranaceis herbaceis lanceolatis basi late attenuatis sessilibus costatis lineaque centrali pallida, soris versus apicem biserialibus costamque approximatis erecto-patentibus linearibus oblongisve.

HAB. Vera Paz, Guatemala, elev. 3500 to 5000 feet, *Osbert Salvin, Esq.*

In color, texture, and a good deal in form, this Fern has a very considerable affinity with our *Antrophyum Galeottii* (Tab. 70); but there are characters present too important to allow the two to be considered identical, whether in respect to species or even Genus. The form is different, exactly lanceolate not at all approaching to lorate (strap-shaped) the texture is much more pellucid so that the *venation* is readily distinguishable (which is so indistinct in the *Antrophyum* as to have escaped the notice of some authors.) We have already, in describing the *Antrophyum Galeottii*, spoken of the frequent presence of short linear or oblong sori in series, as in *Grammitis*: but such are always nearer the margin than the costa, and invariably mixed with branched, if not anastomosing sori; here, on the other hand, the sori are placed nearer the costa than the margin, with great regularity; and though varying in length, always undivided, everything indeed indicating the section *Loxogramme* of the Genus *Grammitis*. This species we name in compliment to Osbert Salvin, Esq., who allowed us to share in a beautiful collection of Ferns he lately collected in Guatemala and Mexico.

TAB. LXXI. Fertile plant of *Grammitis Salvinii*, Hook.: *natural size. Fig.* 1. Portion of a fertile frond, with a linear sorus. *f.* 2. Portion of fertile frond, with an oblong sorus;— *magnified.*

Tab. LXXI

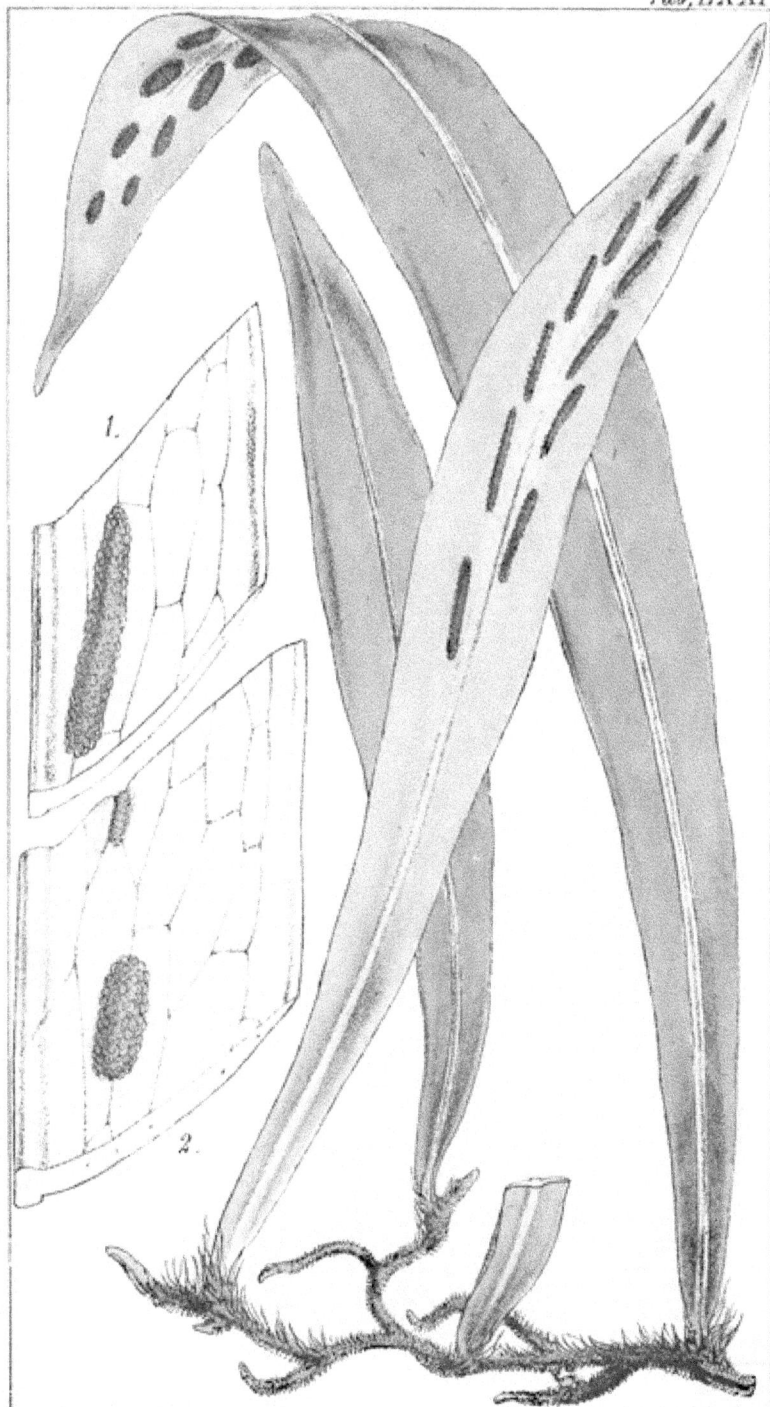

Fitch del et lith.

Pamplin, imp

TAB. LXXII.

CHEILANTHES INTRAMARGINALIS, *Hook.*

Var. *grosse serrata.*

Cheilanthes intramarginalis, *Hook. Sp. Fil.* 2 p. 112, (*which see for description, synonyms, and remarks*). *Metten. Cheilanthes*, p. 49.

Var. Segmentis fertilibus grosse serratis. *Metten.* l. c. p. 50, f. 38-41. (TAB. NOSTR. LXXXII.)

Pteris fallax. *Mart. et Gal. Fil. Mex.* p. 53, t. 14, f. 2.

HAB. Mexico and Guatemala, *Martens et Galeotti*, Volcan de Agua, Guatemala, elev. 6000-7000 feet, *Osbert Salvin, Esq.*

In what has been considered the normal state of this plant, the sterile fronds alone exhibit, and very indistinctly, serratures; the fertile fronds none. But beautiful specimens before us of this species (some of them 18 inches long), in full fructification, have the segments so strongly serrated that I at first looked upon this as a new species, and it certainly is that state of *Cheilanthes intramarginalis*, which Martens and Galeotti published as distinct from that species under the name of *Pteris fallax*. It is a satisfaction to us to find that Mettenius in his recent work on the Genus *Cheilanthes*, agrees with us in referring the plant to *Cheilanthes*, rather than to *Pteris*, or *Allosorus*, or *Pellæa*, or *Cassebeera*, or *Platyloma*, in which several genera it has been placed according to the respective views of authors who have written upon it.

TAB. LXXII. Represents a fertile frond of *Cheilanthes intramarginalis*, var.; *natural size. Fig.* 1. Fertile pinna. *f.* 2. Portion of an involucre; and *f.* 3. Portion of a sterile pinna;—*magnified.*

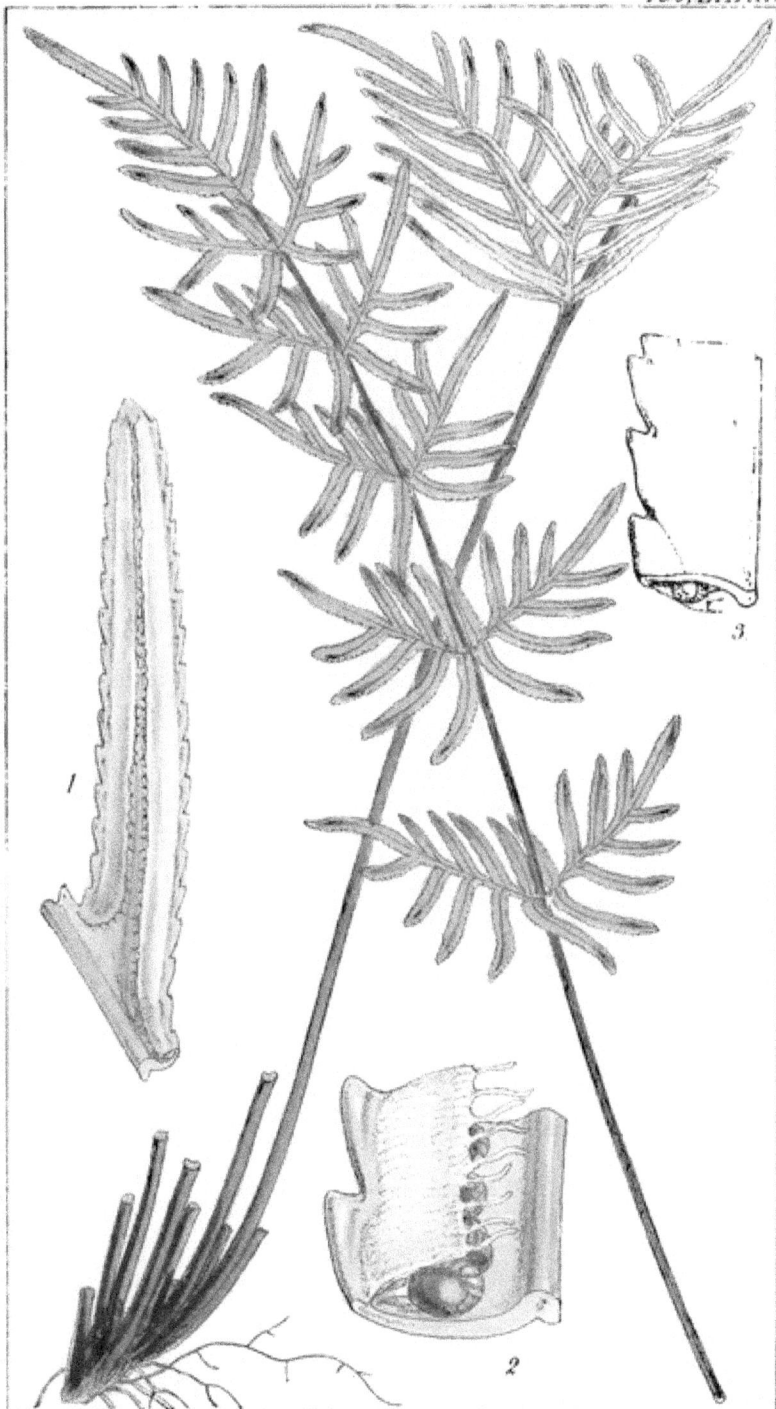

TAB. LXXIII.

ANTROPHYUM MANNIANUM, *Hook.*

Caudice brevi repente dense tomentoso-radiculoso, stipitibus approximatis gracilibus complanatis, frondibus amplis 6-8 uncias longis latisque rhombeo-rotundatis membranaceis firmis subpellucidis (siccitate fusco-olivaceis) caudato-acuminatis subsinuato-serratis basi brevissime attenuatis ecostatis, venis conspicuis elevatis ubique anastomosantibus, areolis oblongis, soris superficialibus sæpe interruptis vel subcontinuis.

HAB. Epiphytal, on trees, Peak of Fernando Po, at an elevation of 3000 feet above the sea level, *Gustav Mann, n.* 367.

This is unquestionably the finest species of the beautiful Genus *Antrophyum* yet known to us: and is one of the many novelties that rewarded our admirable collector, Mr. Gustav Mann, for his late arduous but successful ascent of the famous tropical Peak of Fernando Po, whose elevation is estimated at 10,700 feet. The nearest affinity of the species is, doubtless, with the *A. latifolium*, Blume, Fl. Jav. p. 75, in note, (*A. Boryanum, in the text*, and on the plate, Tab. 31, and of Fée, but not of Kaulfuss, or Hook. et Grev.): but it is truly distinct; Blume's plant being much smaller, of a carnoso-coriaceous, very firm texture, quite opaque when dry, with sunken veins; its colour, when dry, pale yellowish green, so that the copious brown sori which occupy the disk (not extending to the margin) are exceedingly conspicuous on the pale coloured frond: the base is gradually attenuated into the shorter and broader stipites.—In our plant the dry dark-colored frond is so membranaceous and pellucid that the minutely cellular texture is distinctly seen with a magnifying lens of small power, and the venation is very conspicuous, slender, firm, and as it were prominent (not sunk).

TAB. LXXIII. Plant of *Antrophyum Mannianum*, Hook., fertile; *nat size. Fig.* 1. Portion of a fertile frond, with sori; *magnified.*

Tab. LXXIII

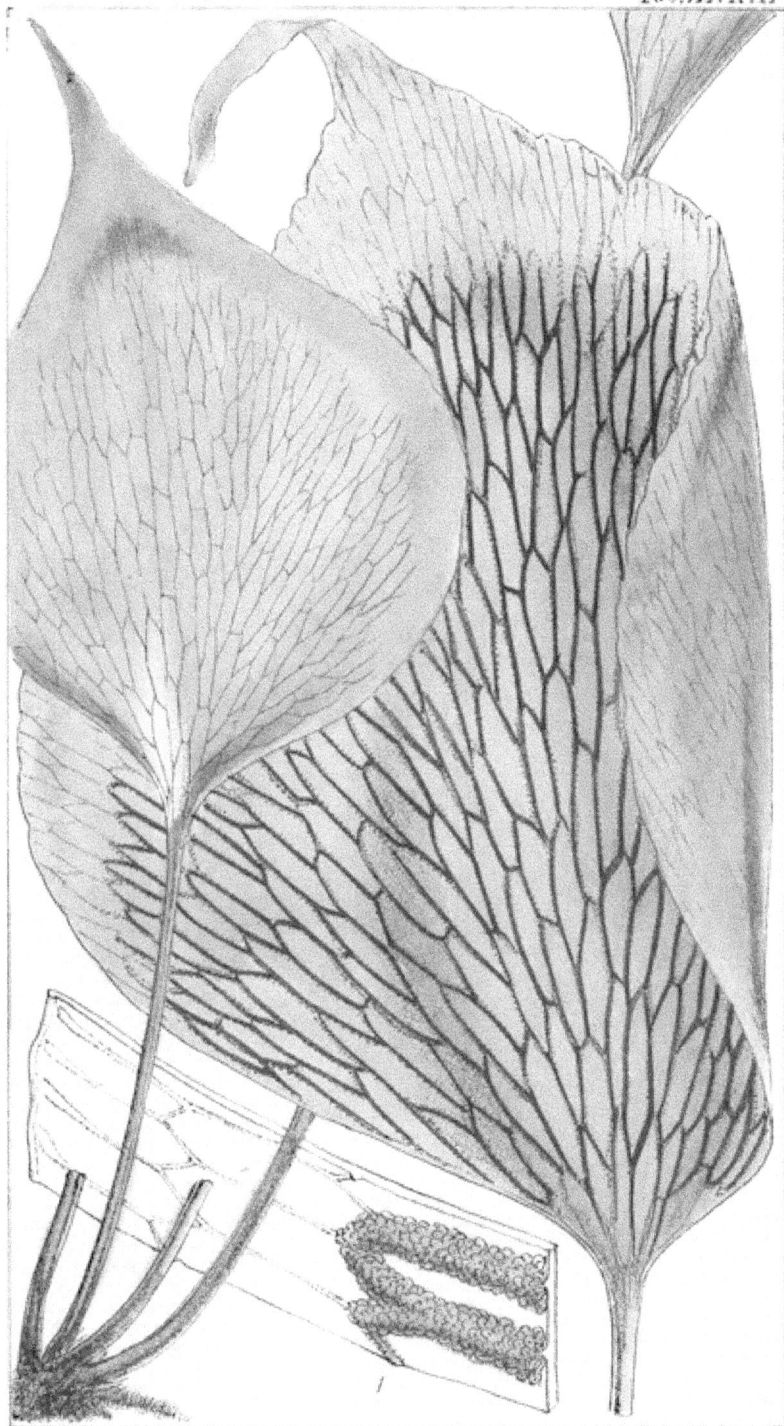

TAB. LXXIV.

EQUISETUM GIGANTEUM, *L.*

Caule erecto stricto 10-14-pedali et ultra diametro unciam sequiunciam arcte striato læviusculo copiose verticillatim ramoso, ramis patentissimis numerosis 6-12 uncias longis gracilibus semilineam ad lineam latis simplicibus vel parce ramulosis asperiusculis, caulis vaginis unciam longis (siccitate pallide testaceis), dentibus subulatis aterrimis magis minusve unitis sæpe semiunciam longis, ramorum dentibus parvis liberis albis rarius atris, amentis ovato-cylindraceis acutis semipollicaribus.

Equisetum giganteum, *Linn. Sp. Pl.* p. 1517, *Willd. Sp. Pl.* 5. p. 9.

Equisetum Poeppigianum, *A. Braun, Mst. in Fil. Lechler.* p. 21, *and in Lechl. Pl. Peruv, n.* 1556, f. 2, *name only.*

Equisetum ramosum, altissimum, *Plumier, Plant. Amer.* 2, p. 115. t. 125.

HAB. West Indies, *Plumier;* Jamaica, *Sloane;* Arica, Peru, *Lechler.*

We are accustomed to see in geological collections fossil specimens of gigantic *European Equiseta,* such as neither Europe nor any part of the old world now possess in a living state: but tropical America affords the present remarkable existing species which almost vies with the fossil forms above alluded to. Plumier has well represented a portion of the plant from Martinique. Sloane, and Patrick Brown, and Lunan record a very large "arborescent" *Equisetum* in Jamaica, no doubt this species. Lechler's specimens are very much broken, for the plant seems very fragile, and our representations are all fragmentary. Lechler found it at Arica in Peru. No author however has made any mention of its height. Our friend, Mr. Spruce, in all probability alludes to this species when writing from the interior of South America, of a gigantic species of *Equisetum,* 20 feet high!

TAB. LXXIV. *Fig.* 1, 2, 3. Portions from different parts of a main stem of *Equisetum giganteum L.;* and *f.* 4. Fertile branch of the same; *natural size. f.* 5. Young spike or amentum of flowers; not yet emerged from its sheath, slightly *magnified. f.* 6. Fully formed spike, ditto; *f.* 7. Front view of a fertile scale of the amentum; *f.* 8. Side view of ditto; and *f.* 9 and 10. Capsules with their clavate, spiral filaments;—*magnified.*

Tab. LXXIV

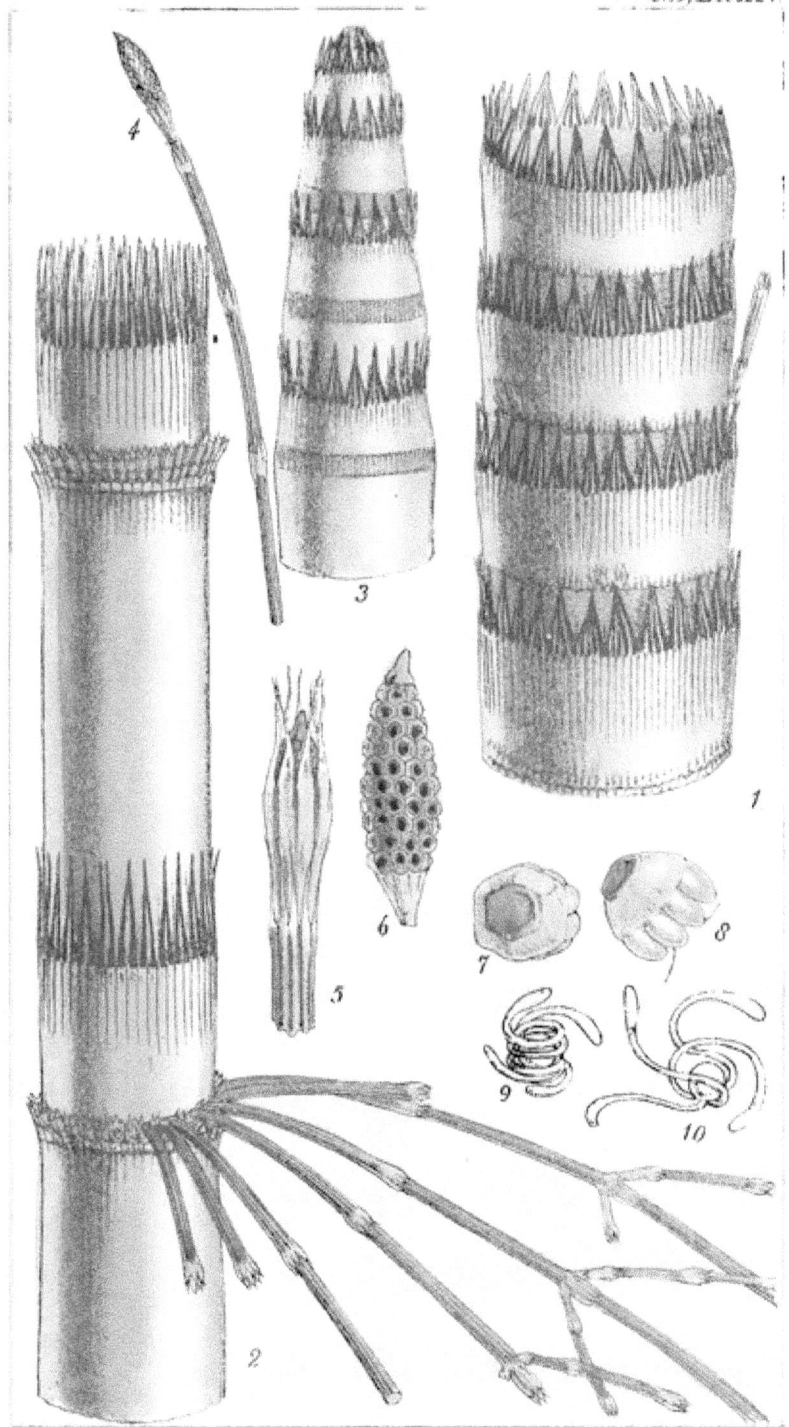

TAB. LXXV.

·

ASPLENIUM (EUASPLENIUM) BARTERI, *Hook.*

Parvum, caudice subnullo, radicibus copiosissimis cæspitosis, stipitibus aggregatis gracilibus 1-2 uncialibus nigro-brunneis, frondibus subquadriuncialibus membranaceis atroviridibus pinnatis, pinnis 30-35 semiunciam longis horizontalibus approximatis nunc omnibus oppositis sessilibus oblongis acutis basi oblique truncatis serratis superne auriculatis, terminali elongata gracili remote pinnatifida sæpe prolifera, venis remotiusculis, soris oblongis, rachi intense fusca compresso-alata.

HAB. Tropical Western Africa; Aboh, on Trees, *Barter*, in Baikie's Niger Exped. *n.* 1454.

This pretty *Asplenium* may rank near *A. pteropus*, Kaulf., and Hook. Sp. Fil. p. 122, t. 177; but it is very different from that and every other of the Genus with which I am acquainted. Instead of tapering gradually to an acuminated apex, the frond is, as it were, suddenly truncated and a terminal pinna set on of a different shape from the lateral ones, and very generally proliferous.

TAB. LXXV. *Asplenium Barteri*, Hook,; *natural size* *Fig.* 1. Fertile pinna; *magnified*; and *f.* 2. Sori, more highly *magnified*.

Tab. LXXV.

TAB. LXXVI.

LYGODIUM (EULYGODIUM) POLYSTACHYUM, *Wall.*

Villosulum, longe scandens, ramis bipinnatis, pinnis remotis geminatis 1-1½ pedalibus ad basin sæpe gemmiferis, pinnulis alternis remotis petiolulatis in petiolulum articulatis, sterilibus fertilibus conformibns oblongis vel ovato-lanceolatis pinnatifidis, segmentis oblongis obtusis, fertilium laciniis soriferis, rachibus pubescentibus hinc longeque ferrugineovillosis, venis liberis.

Lygodium polystachyum, *Wall. Cat.* n. 2200.

HAB. Malayan Archipelago and Peninsula. Woody mountains of Pulo-Penang, *Wallich.* Mergui (*Griffith*) and Tonglow, and Mergui, *Rev. C. S. P. Parish*, n. 46, Moulmaine, *Thos. Lobb.*

An undescribed species of *Lygodium*, and unquestionably among the most distinct of the Genus. Dr. Wallich was the first to discover it, and we have specimens also from Griffith, and from Mr. Parish. It appears to be an extensive climber, the main rachis is deciduously villous and pubescent. The primary pinnæ spring in pairs from one point or side of the main rachis, and often bear a shaggy gemma between them at the base. The pinnæ are 2-2½ inches long, and what is, as far as I know, peculiar to this species, the fertile pinnæ do not essentially differ from the sterile ones, they are merely a little narrower, and the segments, somewhat contracted, bear the two rows of capsules on the underside.

TAB. LXXVI. *Fig.* 1. Portion of a fertile plant, of *Lygodium polystachyum*, Wall.; *natural size. f.* 2. Sterile pinna; *nat. size. f.* 3. Sterile segment; *magnified. f.* 4. Fertile segment; *magnified. f.* 5. Sorus; and *f.* 6. Capsule; *more magnified.*

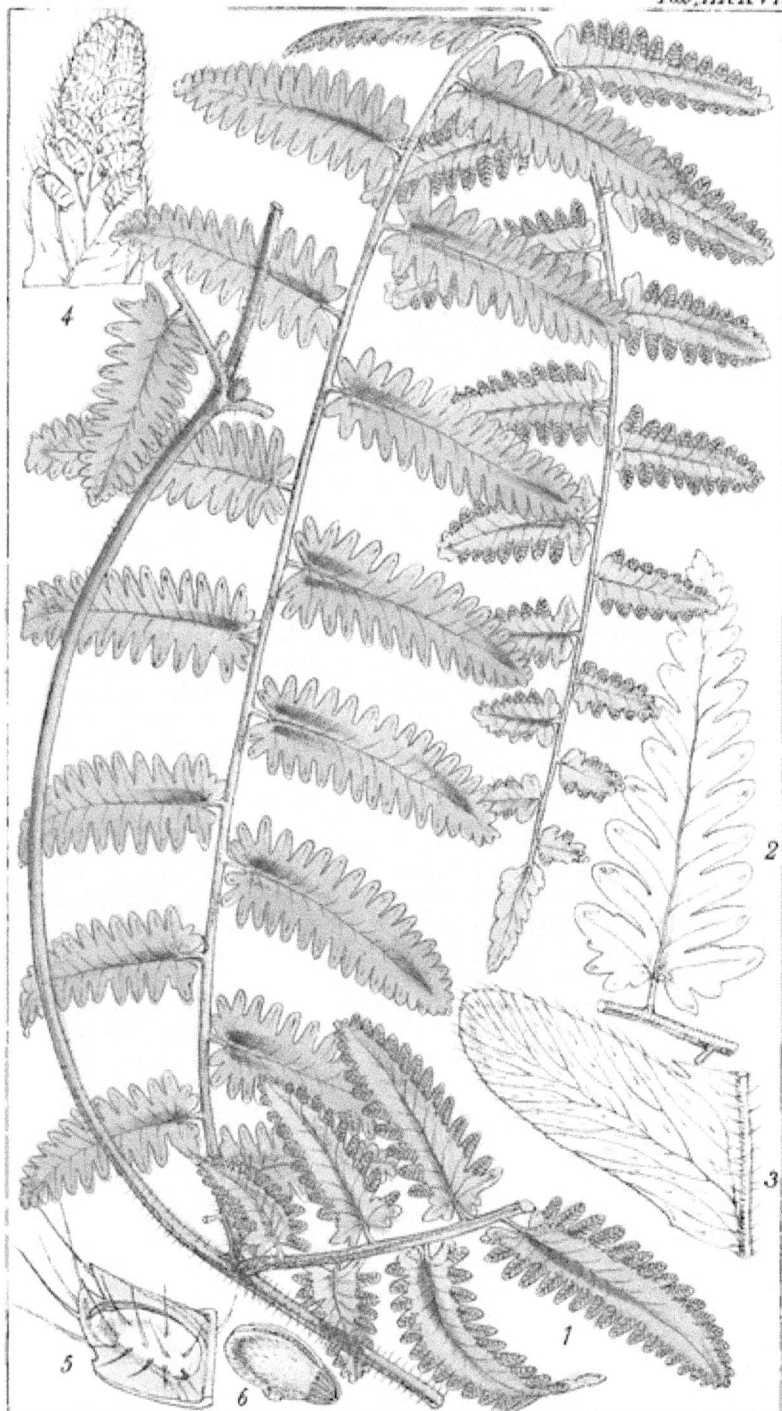

Tab. LXXVI.

TAB. LXXVII.

Nothochlæna Rawsoni, *Pappe.*

Caudice longe repente squamis membranaceis rigidis subulatis nigro - costatis paleaceo, stipitibus approximatis demum nudis purpureo-ebeneis, frondibus linearibus spithamæis pinnatis, pinnis alternis remotiusculis subsessilibus cordato-ovatis obtusis crasso-coriaceis lobato-pinnatifidis supra nudis subtus densissime albido-v.-ferrugineo-pannoso-villosis, lobis 5-9 rotundatis obtusis, marginibus subincrassatis patentibus, soris marginalibus continuis, capsulis nigris.

Nothochlæna Rawsoni, *Pappe, in Pappe and Rawson's Syn. Fil. Afr. Austr.* p. 42.

Hab. On hills between Spectakel and Komaggas, Namaqualand, S. Africa, *Rev. H. Whitehead,* 1856.

Much as this is allied in habit and general appearance to the South American *Nothochlæna rufa* (see Tab. LII. of this volume) and to some other species of that Genus, and of *Cheilanthes* of S. America, it is in reality extremely different. The caudex is long and creeping, clothed with imbricating, subulate, erose scales, which are very rigid in consequence of the broad black costa which runs through the centre. The pinnæ are green, and free from tomentum above, beneath densely woolly, sometimes rich tawny or ferruginous, sometimes nearly white. The margin is not reflexed, nor at all involuciform, so that the black capsules are quite exposed, forming a continuous line along the margin.

This is well described by Pappe and Rawson in their Synopsis of the Ferns of S. Africa, and is very rare in the Colony, only one locality having yet been discovered, and that was detected by the *Rev. Mr. Whitehead.* I am indebted to Wm. Rawson, Esq. C.B., and to Rear-Admiral Sir Frederick Grey, K.C.B., for fine specimens.

Tab. LXXVII. Plant with sterile and fertile fronds of *Nothochlæna Rawsoni,* Pappe; *natural size. Fig* 1. Scale from the caudex; *f.* 2. Upper side of a pinna; *f.* 3. Under side of a fertile pinna; *f.* 4. Portion of sorus; *f.* 5. Capsule; *f.* 6. Hairs from the tomentum at the back of the frond; *all more or less magnified.*

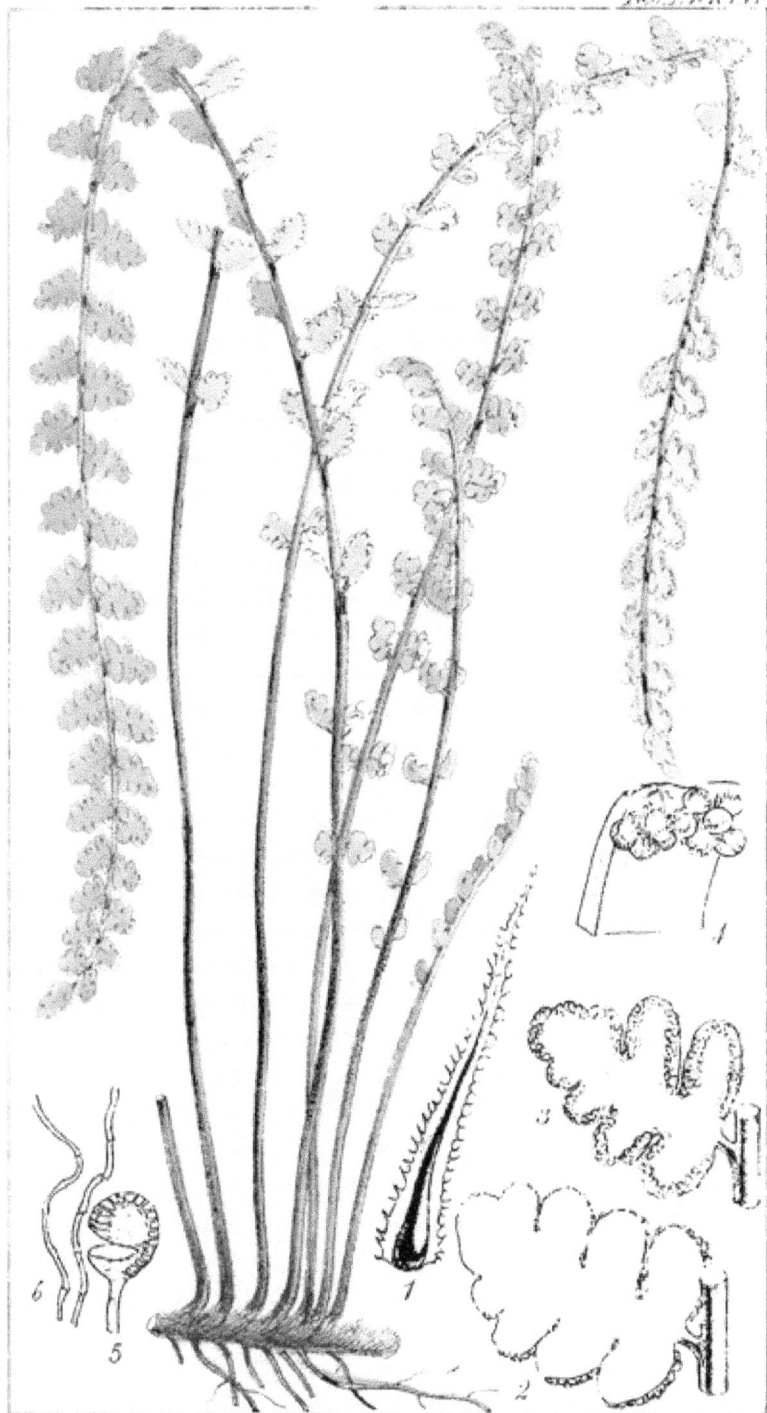

Tab. LXXVII

TAB. LXXVIII.

GYMNOPTERIS (LEPTOCHILUS) MINOR, *Hook.*

Parva, caudice repente crassitie pennæ anserinæ apice squamoso, stipitibus remotis gracilibus filiformibus, frondibus
sterilibus membranaceis oblongis lato-lanceolatisve costatis
in stipitem 2-2½ uncias longum attenuatis, venis anastomosantibus angulato-areolatis, areolis appendiculatis prope
marginem minoribus, venulis ultimis clavatis liberis; frondibus *fertilibus* biuncialibus linearibus in stipitem quatuor
uncias longum attenuatis.

Leptochilus minor, *Fée, Acrost.* p. 87, t. 25, f. 87 (*excl. the
synonym of* Gymnopteris normalis, *J. Sm.*)

HAB. Subtropical region, Khasia, near Churra, elev. 2-3000
feet; *Hook. fil. et Thomson.* Isle of Samar, Phillipines,
Cuming, n. 326 (*according to Fée*).

That this is the *Leptochilus minor* of Fée, there can be no
doubt; and that author seems to have taken his figure from
Cuming's Phillipine Island plant, n. 326 (the only locality
he gives for the species), and he quotes J. Smith's *Gymnopteris normalis,* a name, without description unfortunately, to
Cuming's n. 326; but that number in my Herbarium has
quite, or nearly quite, sessile sterile fronds, whereas Fée's
plant he figures and describes "frondibus sterilibus longe
petiolatis:" so that it would appear that two species have
been distributed by Mr. Cuming under the same number.
The present one is remarkable for its small size, and the
great comparative length and slenderness of the stipites;
especially of the fertile ones.

TAB. LXXVIII. Plants of *Gymnopteris minor,* Hook.;
natural size. Fig. 1. Portion of a sterile frond; and *f.* 2.
Portion of a fertile frond; *magnified.*

Tab. LXXVIII.

TAB. LXXIX.

Antrophyum Brookei, *Hook.*

Caudice repente subulato-squamoso densissime olivaceo-
tomentoso, frondibus cæspitosis membranaceis flaccidis 3-
pollicaribus ad spithamæam lineari-lanceolatis acuminatis
basi attenuatis sessilibus ecostatis, soris anguste linearibus
2-4 sæpe longe continuis simplicibus nunc interruptis et
parce ramosis immersis, venis anastomossantibus areolas
valde elongatas margini parallelas formantibus, capsulis pilis
articulatis intermixtis.

Hab. On trees, Sarawak, Borneo, *Thos. Lobb.* Naviti Levu,
Fiji Islands, on mountains, *Milne,* in Voy. of H.M.S. Herald.
Samoan Islands, *Rev. Mr. Parker.*

The caudex of this is creeping, but as well as the roots, often
densely covered with a mass of olivaceous tomentum, so as
scarcely to be visible. Where the tufts of fronds arise the
caudex is seen to be paleaceous with subulate scales. Fronds
membranaceous, flaccid, three inches to a span, or almost a
foot long, one-third of an inch wide in the broadest part,
narrow, linear-lanceolate, accuminated, ecostate, much tapering below, but not stipitate; a darkish line indeed runs through
the very narrow base, but nothing that can be called a mid-
rib. The sori vary in form. In one specimen two uninter-
rupted longitudinal lines run for a length of five inches between
the centre and the margins, and a third but shorter continuous line appears between one of these and the margin; in
other specimens the sori form two to four somewhat parallel
lines, here and there branched and variously interrupted; all
are sunk in a channel or groove in the substance of the frond,
and all arise from the veins which run longitudinally, while
oblique veinlets unite them so as to form very elongated
areoles.

I can no where find any Fern described corresponding
with this. It seems to approach the *Antroph. angustatum* of
Brackenridge from Tahiti; but that has a stipes four inches
long. It very much resembles in general aspect the *Antroph.
lineatum,* Kaulf. (*Polytaenium lineatum, Desv.* and *Hook.
Gen. Fil.* Tab. CVII.), but that has copious parallel sori and a
distinct costa. Our *Antrophyum Galeottii* (see Tab. LXX.
of this volume), has the areoles and sori oblique.

Tab. LXXIX. *Fig.* 1 & 3. *Antrophyum Brookei,* Hook.;
fertile plants; *natural size.* *f.* 2. Portion of a frond, showing
the venation. *f.* 3. Grooved receptacle of the capsules.
f. 4. Capsule and accompanying articulated hairs; *magnified.*

Cent. 2. t. 79.

Tab. LXXIX.

TAB. LXXX.

ASPLENIUM (EUASPLENIUM) EMARGINATUM, *Beauv.*

Caudice brevi crasso erecto dense fibroso-radicoso, stipitibus
aggregatis rachique subherbaceis, fronde ampla pedali ad
sesquipedalem submembranacea pinnata læte-viridi, pinnis
4-5-pollicaribus sesquiunciam et ultra latis brevi-petiolatis
oblongis obtusis dentato-serratis basi oblique cuneatis apice
profunde acute emarginatis sinu gemmifera, terminali majore longe petiolata, venis patentibus copiosis uni-bifurcatis,
soris numerosis approximatis costam non ad marginem approximatis, involucris angustissimis albidis.

Asplenium emarginatum, *Beauv. Fl. d'Oware et de Benin,* 2,
p. 6, t. 61. *Metten. Asplen.* p. 94, *Hook. Sp. Fil.* 3, p.
101 (in part).

HAB. Tropical Western Africa, Mountains of Isle du Prince,
Bight of Benin, *Palisot de Beauvois;* Onitoba, *Barter,* in
Baikie's Niger Expedition, n. 1735. Fernando Po, on
mountains 1000 feet of elev. *Gustav Mann,* n. 343.

I am led to believe, in consequence of more perfect specimens I have lately received of Beauvois' *Asplenium emarginatum,* that Dr. Curror's specimens alluded to in my habitats
for that plant in the Species Filicum, are different from M.
de Beauvois', and I gladly correct my error, by publishing at
our Tab. LXIX. of this volume, Dr. Curror's plant under
the name of *A. longicauda,* while I here represent what I
believe to be quite different, and the true plant of the author
of the " Flora d'Oware et de Benin." It will be at once
seen that the present plant entirely wants the suddenly
accuminated points to the apices of the former species, and
equally the very long proliferous terminal pinna : here all
the pinnæ are obtuse and emarginate. The texture of the
frond is more membranaceous, of a brighter green, and the
veins are more compact, the sori much closer, and longer and
narrower.

TAB. LXXX. Frond of *Asplenium emarginatum,* Beauv.;
natural size. Fig. 1. Portion of a fertile pinna, with sori ;
magnified. f. 2. Two lateral pinnæ, fertile ; *natural size.*

Tab. LXXX.

TAB. LXXXI.

CHEILANTHES KIRKII, *Hook.*

Caudice ascendente radiculoso crasso, stipitibus spithamæis et
ultra cæspitosis rigidis atro-purpureo-ebeneis nitidis inferne
subulato-squamosis, frondibus 4-pollicaribus coriaceis opacis
cordiformibus profundissime tripartitis 5-lobo-palmatis, divisionibus primariis infimis semitriangularibus intermedia
triangulari omnibus bipinnatifidis laciniis ultimis oblongo-
lanceolatis acutiusculis sinubus acutis, soris copiosis uniformibus, involucris subrotundis reniformibusve pallide fuscis
venas furcatas terminantibus, costis subtus aterrimis nitidis.

HAB. Moramballa Mountain, Zambesi, elev. 3000-3500 feet.
Dr. Kirk, in *Dr. Livingstone's Zambesi Expedition, Dec.*
1858.

This very interesting plant has so entirely the habit and
general structure of the well known *Pteris,* or *Pellea* as it is
now by many called, *geraniifolia,* (see first Cent. of Ferns,
Tab. 15), that, without fructification, I cannot point a single
character by which the one can be distinguished from the
other; but the sori of our present plant, are everywhere, and
upon all our specimens, so entirely these of *Cheilanthes,* that, so
long as that genus retains a place in our system, this plant
must be referred to it. It is true we have had occasion to
remark of several species of *Cheilanthes,* such a degree of
confluence in the sori as to render it doubtful whether they
should belong to one genus or the other, but here we have in
the sori of *Pellea geraniifolia* and of *Cheilanthes Kirkii* the
extremes of the two kinds of fructification, uniformly *distinct*
in the one, and quite *continuous* in the other.

TAB. LXXXI. Plant of *Cheilanthes Kirkii,* Hook.; *natural size. Fig.* 1. Fertile segment of a frond; *magnified; f.*
2. Portion of a segment with sori, and showing the venation;
more magnified.

Tab.LXXXXI

TAB. LXXXII.

ASPLENIUM (EUASPLENIUM) SEELOSII, *Leyb.*

Nanum, frondibus longe stipitatis trifoliolatis, foliolis sessilibus lanceolatis integerrimis vel grosse serratis undique distincte hirsutis pilis diaphanis, involucris albidis marginibus suberosis approximatis subimbricatis.

Asplenium Seelosii, *Leybold, in Flora*, 1855, p. 81, and 348, Tab. 15, *Metten. Asplen.* p. 142. *Hook. Gen. et Sp. Fil.* v. 3, p. 176.

Acropteris Seelosii, *Henfl. Aspl. Europ.* p. 111.

Asplenium tridactylites, *Bartl. in Herb. Kze. and Bolle in Herb. Hook.* (sterile fronds, simple, three lobed).

HAB. Clefts of Dolomitic rocks, at an elevation of from 4-6000 feet, on the south and north-west side of the Schleerngeberg, in South Tyrol, *Seelos, C. Bolle.*

This is perhaps the rarest of all European Ferns, confined to a very limited locality in South Tyrol. The affinity is evidently with small specimens of *Aspl. septentrionale* or *Aspl. Germanicum*, and hence the plant would be an *Acropteris* of Link. The only specimens I possess were obligingly sent to me by *Dr. Bolle;* they were accompanied with the locality " prés de Salurn, au pied du Mont Geier, dans les fentes des rochers calcaires, Tyrol meridionale."

TAB. LXXXII. *Fig.* 1, 2, & 6. Plants of *Asplenium Seelosii*, Leyb. fertile and sterile; *natural size. f.* 4. Perfect state of the fertile frond, upper side; *f.* 5. Fertile pinna, seen from beneath. *f.* 3. Portion of a fertile pinna, with two sori; *all more or less magnified.*

Tab. LXXXII

TAB. LXXXIII.

SELAGINELLA SPRUCEI, *Hook.*

Semipedalis ad pedalem, caule inferne repente stolonifero
superne ascendente pinnatim ramoso, ramis seu pinnis
elongatis arcte foliosis obtusis 6-9 lineas latis, foliis horizontalibus lingulato-oblongis obtusis uninerviis minute serrulatis basi superiore dilatato-rotundatis, stipulis triplo
minoribus ovatis acuminatis subfalcatis erectis imbricatis,
amentis copiosis omnino lateralibus unciam sesquiunciam
longis flexuosis sutetragonis sæpissime deflexis, bracteis
ovatis subacuminatis paululum carinatis.

HAB. On a mountain called Campana, near Tarapoto, Eastern
Peru, *R. Spruce,* n. 4623.

This, I have reason to believe, is quite new, and certainly
a most lovely species, worthy of bearing the name of so distinguished a traveller and so able a Botanist, as its discoverer,
Mr. Richard Spruce. Its nearest affinity is perhaps with
S. Breynii, Spring.

TAB. LXXXIII. Upper portion of a plant of *Selaginella
Sprucei,* Hook.; *nat. size.* *Fig.* 1. Upper side of a portion
of a branch, with leaves, stipules and a spike or amentum;
magnified. *f.* 2. Leaf; *more magnified.* *f.* 3. Bractea and
capsule, and *f.* 4. Spores; *more magnified.*

Tab. LXXXIII

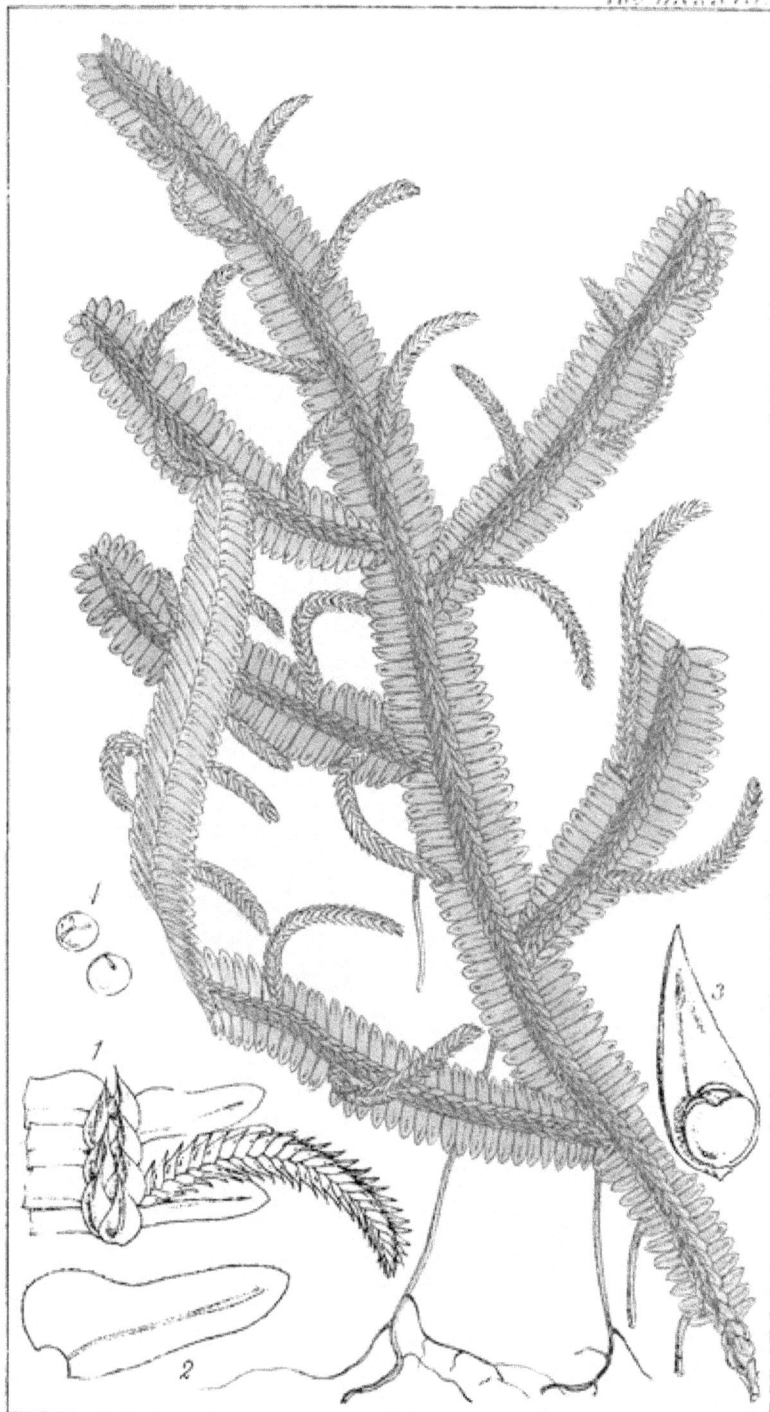

TAB. LXXXIV.

SELAGINELLA SUBARBORESCENS, *Hook.*

Caule sesquipedali erecto terete parce et appresse folioso basi radicoso, superne frondoso pinnatim et dichotome ramoso ; fronde coriacea erecta 12-16 uncias longa, pinnulis vel ramulis 6-8 uncialibus semiunciam latis linearibus acuminatis apice fertilibus subtus pallidioribus; foliis exacte horizontalibus oblongis acuminato-acutis paululum falcatis integerrimis uninerviis, stipulis parvis erectis imbricatis, amentis terminalibus parvis binis ternisve lineari-oblongis nunc subramosis; bracteis cordato-ovatis acuminatis subserratis.

HAB. Rio Uapés, tributary of the Amazon, Brazil, in a forest near Jauaraté-Cochocira, *R. Spruce, n.* 2540.

This noble species is, perhaps, the tallest of the genus, the largest specimens that came under Mr. Spruce's notice, he observes, were 4½ feet high : so that I have thought the name of *subarborescens* not inappropriate. The leaves on the stipes are close pressed, dark green, ovate, very distant, upwards they gradually, as it were, pass into the foliage of the frondose portion. The stipules are remarkably small in proportion to the size of the leaves.

TAB. LXXXIV. Stipes and portion of a frond of *Selaginella subarborescens,* Hook. ;—*natural size. Fig.* 1. Stipules and leaves, seen from above; *f.* 2. The same seen from beneath ;—*magnified. f.* 3. Capsule ; *f.* 4. One of the three seeds from the capsule ; *f.* 5. Bractea and antheridium, and *f.* 4. Pollen grain ; *all more highly magnified.*

Tab. LXXXIV

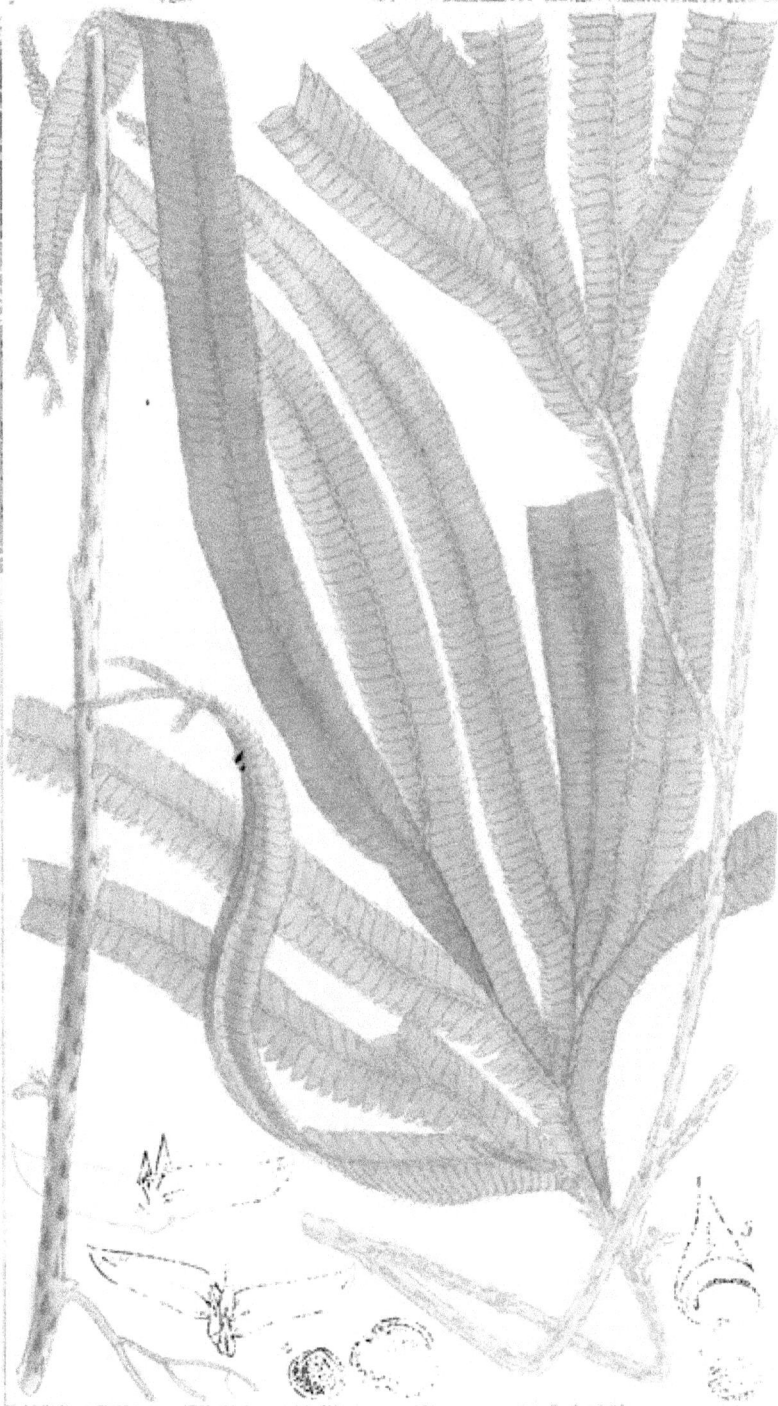

TAB. LXXXV.

SELAGINELLA DENSIFOLIA, *Spruce.*

Quadri-sexuncialis, procumbens, subcoriacea, nitidissima (subtus præcipue) bipinnatim ramosa copiose flagellifero-radi-
cans, ramis lato-linearibus siccitate supra canaliculatis subtus convexis 1½-2 lin. latis apice fructiferis, foliis dense
imbricatis patentibus oblongis obtusis inæquilateralibus
uninerviis serratis inferne ciliatis basi superiore dilatatis
rotundatis, stipulis parvulis ovatis acuminatis imbricatis,
amentis terminalibus solitariis vel binis semiunciam longis,
bracteis imbricatis ovatis acuminatis serratis.

Selaginella densifolia, *Spruce, mst. in Herb. nostr.*

HAB. Cerro de Morro, in shady moist places of the River
Orinoco, *R. Spruce,* n. 3809.

A small but extremely elegant species, remarkable for its
very compact, densely imbricated and glossy foliage, dark
green above, pale beneath; and for the copious flagelliform
roots, which descend from the underside of the plant.

TAB. LXXXV. Plant of *Selaginella densifolia,* Spruce;
natural size. Fig. 1. Apex of a branch with its amentum.
f. 2. Stipules and leaves seen from above; and *f.* 3. Leaves
seen from beneath; *magnified. f.* 4. Single leaf, and *f.* 5.
Bractea, with antheridium;—*more magnified.*

Tab. LXXXV.

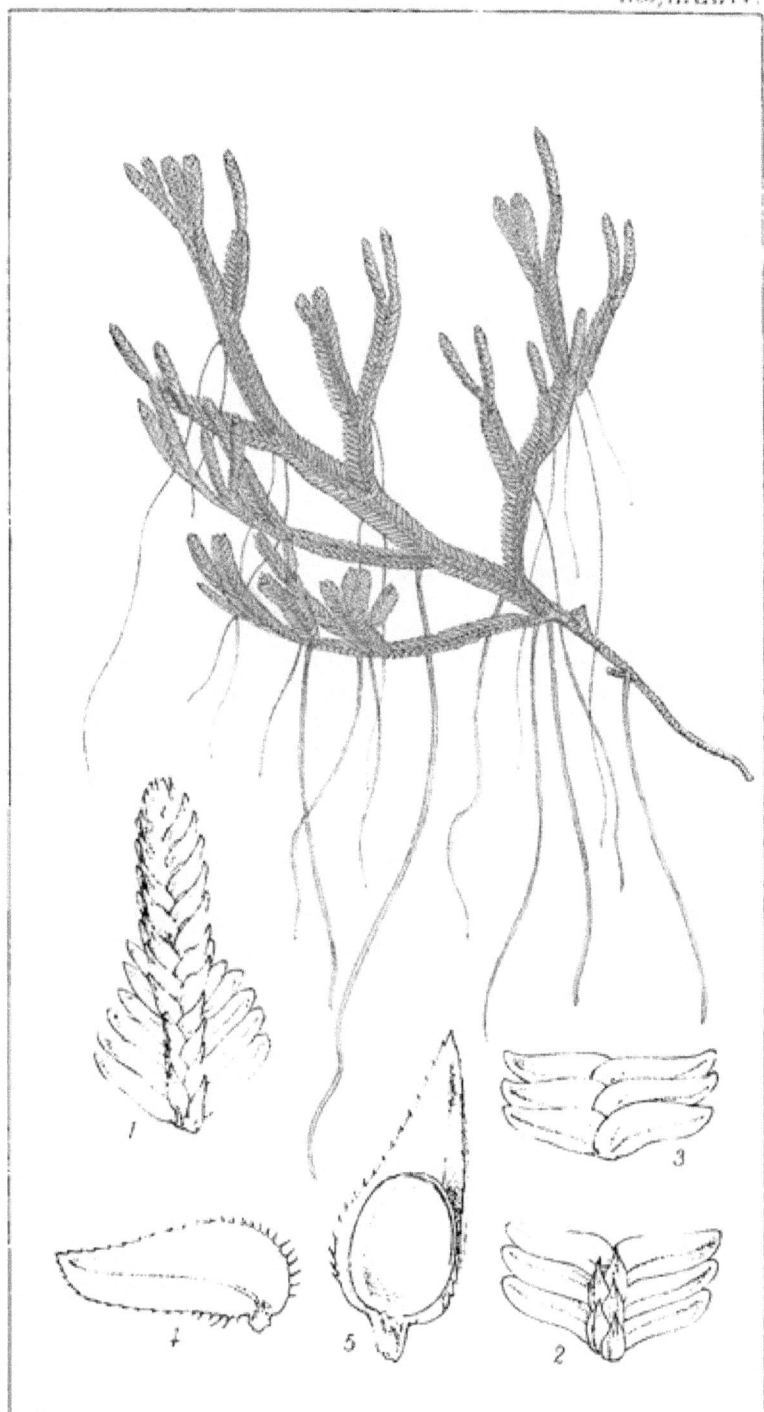

Fitch del. et lith.   Pamplin imp.

TAB. LXXXVI.

SELAGINELLA VOGELII, *Spring.*

Bi-tripedalis, caudice repente, stipite pedali ad sesquipedalem erecto nudiusculo ramisque primariis stramineo-fusco nitido, fronde stipitis longitudine orbiculari-ovata 3-4-pinnata, pinnis ultimis oblongo-linearibus, foliis horizontaliter patentibus distichis subcoriaceis subsesquilineam longis, primariis longe distantibus ovato-acuminatis, reliquis arcte approximatis vix imbricatis lineari-oblongis acutis integerrimis uninervibus subtus glaucis basi inferiore adnata superiore rotundata-dilatata; stipulis 4-plo minoribus ovatis longe acuminatis imbricatis, amentis linearibus terminalibus subtetragonis laxis, bracteis ovato-acuminatis subcarinatis erecto-patentibus.

Selaginella Vogelii, *Spring Monogr. Lycop.* p. 171.

HAB. Fernando Po. *Vogel, Barter*, n. 1044, 1398; *Mr. Mann,* n. 149; Isle of Nissobe, East Coast of S. Africa, *Boivin.*

Linnæus in his Species Plantarum enumerated twenty-four species of *Lycopodium.* The Genus was subsequently properly divided into two, *Lycopodium* and *Selaginella,* and the number was increased by Spring, about 20 years ago, to 101 species of *Lycopodium,* and 209 of *Selaginella.* They are plants of great beauty, the latter genus especially, but notoriously difficult to be clearly discriminated and characterized, and hence with some reason they have been called the "indeterminable *Lycopodiaceæ.*" Our own Herbarium would now add considerably to the number of species we originally contributed to Dr. Spring's Monograph—on the other hand we fear that this able author has needlessly multiplied species. It may be some assistance to future Monographists to give accurate figures. The present is a most lovely plant and apparently common in Fernando Po, and as far as we know is only found in that island, and in one spot off the East Coast of Africa.

TAB. LXXXVI. Caudex, stipes, and portion of a primary branch of *Selaginella Vogelii,* Spring ;—*natural size. Fig.* 1. Upper side of leaves and stipules : *f.* 2. Under side of ditto ; *f.* 3. Portion of a fertile spike or amentum ; *f.* 4. Bractea and capsule, and *f.* 5. Capsule ; *all magnified.*

Tab. LXXXVI

TAB. LXXXVII.

GYMNOPTERIS (LEPTOCHILUS) PANDURIFOLIA, *Hook.*

Caudice repente crassitie pennæ anserinæ copiose squamoso
subtus valde radicante, stipitibus approximatis robustis
squamis fuscis squarrosis; *frondium sterilium* stipite semi-
pedali, lamina subpedali membranacea siccitate atro-viridi
late ovato-oblonga panduriformi acuta, basi cordata rotundata lobata pennivenia, venis primariis flexuosis, reliquis
omnibus reticulatis angulatim areolatis, areolis venulis
clavatis ramosis appendiculatis; *frondium fertilium* stipite
pedali, lamina quadriunciali lineari-subpanduriformi.

HAB. Mount Guayrapurima, near Tarapota, Eastern Peru,
R. Spruce, n. 4741.

This would fall into that group of *Gymnopteris* with simple
fronds, which has been distinguished by some authors as a
Genus under the name of *Leptochilus*, but for which I can
detect no valid character. The present, however, is quite
distinct from any described species.

TAB. LXXXVII. Fertile and barren fronds of *Gymnopteris panduriformis*, Hook; *natural size. Fig.* 1. Portion of a
sterile frond, to show the venation; *magnified. f.* 2. Portion
of the fertile frond; *magnified.*

CENT. 2. T. 87.

Tab. LXXXVII.

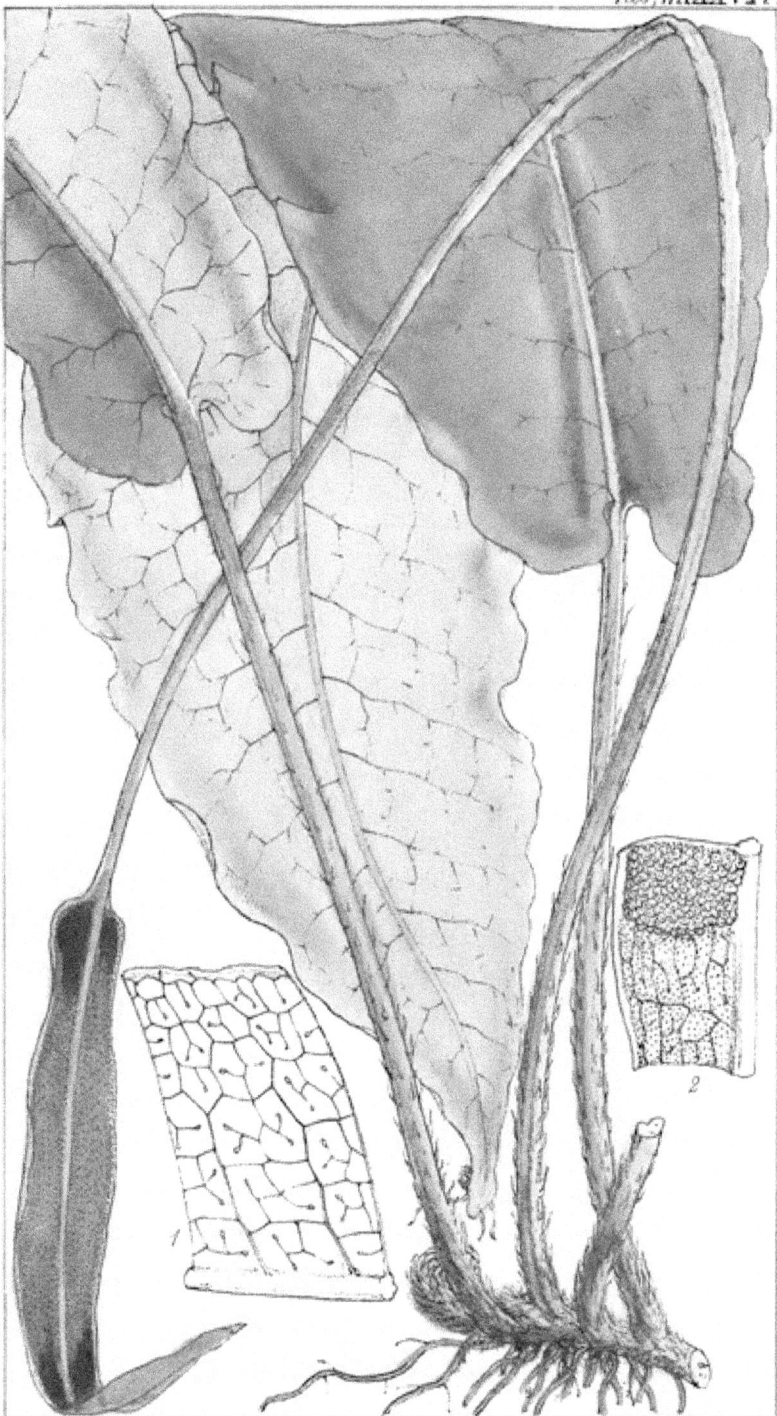

TAB. LXXXVIII.

Acrostichum (Poecilopteris) virens, *Wall.*

Var. minus, fuscatum.

Caudice crasso horizontali, stipitibus approximatis spithamæis
et ultra stramineis inferne sparse squamosis, frondibus 1-2-
pedalibus pinnatis; pinnis *sterilium* 9 ad 20 subsessilibus
oblongo-lanceolatis subduplicato-serratis coriaceo-membranaceis sæpissime læte-virentibus, venis primariis pinnatis
secundariis transversis arcuatis angulato-flexuosis, areolis
sæpe appendiculatis, costularibus majoribus inappendicula-
tis; *fertilium* pinnis linearibus.

Acrostichum virens, *Wall. Cat.* n. 1033, *Hook. et Grev. Ic.
Fil.* Tab. 221.

Campium virens, *Pr. Tent. Pt.* p. 239, *Pr. Epim. Bot.* p. 170.

Cyrtogonium virens, *J. Sm. in Hook. Journ. Bot.* 4, p. 154.

Var. *minus;* fuscatum, pinnis minoribus terminali prolifera,
costis venisque primariis gracilioribus minus conspicuis.

Hab. *Var.* minus, fuscatum. India; Concan, *Mr. Law.* Nilghiri, on stones by the side of streams, Koondah's Corn,
elev. 6000 feet, *McIvor,* n. 4.

Notwithstanding the valuable labours of M. Fée and his
excellent writings and figures of the *Acrostichaceæ,* no group
of Ferns requires more careful study and revision than this
does. Species are, assuredly, too much multiplied, I fear on
very slight grounds : and the multitude of Genera only serve
to puzzle and perplex the student as well as the practical
Pterodologist himself. By the section or subgenus here
called "*Poecilopteris*" we mean that of Presl : but unfortunately, *Poecilopteris* of Presl, adopted by Moore as a Genus,
is *Bolbitis* of Schott, *Campium* of Presl, *Cyrtogonium* of J. Sm.
Heteroneuron of Fée, *Acrostichum* of the older authors : and
if we look to the figures of the generic characters in Mr.
Moore's "Index Fil.," Plate VII. we shall find venation of
two very different kinds; thus allowing a latitude of structure
not tolerated in other Fern genera.

The *Acrostichum* now figured I quite believe may safely be
referred to the *Acrost. virens,* Wall., but the strongly marked
primary, parallel veins here give place to others which gradually merge into an irregular anastomosing as they recede from
the costa ; and, at the margin, is a set of clavate, free veinlets.
The colour and size of the pinnæ are extremely variable, and
I fear that several described species will have to be considered
forms of *A. virens* of Wallich.

Tab. LXXXVIII. Fertile plant of *Acrostichum (Poecilopteris)
virens,* Wall ; *natural size. Fig.* 1. Portion of a sterile pinna, showing the venation ; and *f.* 2, Portion of fertile ditto ; *magnified.*

Tab. LXXXVIII.

TAB. LXXXIX.

LOMARIA (PLAGIOGYRIA) EUPHLEBIA, *Kze.*

Caudice crasso lignoso pedali, frondibus in stipitibus elongatis triquetris basi incrassatis 2-3-verrucosis, 1-2-pedalibus usque ad apicem pinnatis subchartaceis olivaceo-fuscis; *sterilibus* lato-ovatis lanceolatis, pinnis patentibus 5-6 uncialibus remotiusculis elongato-lanceolatis sessilibus acuminatis subserratis basi eglandulosis, venis subdistantibus furcatis; *fertilibus* angustioribus magis oblongis, pinnis linearibus elongatis obtusis, involucro dentato-lacerato.

Lomaria euphlebia, *Kze. in Bot. Zeit.* 6, p 521. *Schk. Fil. Supp.* p. 61, t. 125. *Hook. Sp. Fil.* 3, p. 20.

Acrostichum triquetrum, *Wall. Cat. n.* 25, (in part).

Plagiogyria triquetra, and Pl. euphlebia, *Metten. Plagiog.* p. 10.

Olfersia triquetra, *Pr. Tent. Pterid.* p. 234.

Stenochlæna triquetra, *J. Sm. in Hook. Journ. Bot.* 4, p. 149, *Pr. Epimel. Bot.* p. 165.

HAB. Nepal, *Wallich.* Assam, *Griffith.* Khasia, temperate region, alt. 6000 feet, *Hooker and Thomson.* Tsus Sima, island off the coast of Corea, in lat. 34° ½, *Wilford, n.* 874. Japan, *Goring* (probably quite in the south).

This belongs to an interesting group of *Lomaria,* which the excellent Mettenius has thought worthy to constitute a distinct Genus to which he has given the name of *Plagiogyria.* My reasons for preserving it in *Lomaria* are fully given in the third volume of my Species Filicum, v. 3, p. 2. The present one was long known in our Herbaria as the *Lomaria triquetra,* from Nepal, of Dr. Wallich's Mst. Catalogue. At length it was figured and described as a Japan Fern by Kunze, above quoted. Drs. Hooker and Thomson met with it, but apparently of rare occurrence, in Khasia, Griffith in Assam, and now recently Mr. Wilford sent home specimens from Tsus-sima:—thus it has a very considerable geographical range.

TAB. LXXXIX. Fertile plant of *Lomaria (Plagiogyria) euphlebia,* Kze.; *natural size.* Fig. 1. Portion of a sterile pinna, showing the venation; and *f.* 2. Portion of a fertile pinna; *magnified.*

Tab. LXXXIX.

TAB. XC.

ACROSTICHUM (ELAPHOGLOSSUM) DIMORPHUM, *Hk. et Gr.*

Caudice horizontali vel ascendente crassiusculo radicante apice imbricatim paleaceo, stipitibus copiosis approximatis 4 uncias ad spithamæam longis subrobustis per totam longitudinem sparse paleaceo-squamosis, frondibus *sterilibus* 3-4-uncialibus oblongo-lanceolatis costatis obtusis coriaceo-membranaceis margine lobato-pinnatifidis junioribus minute squamulosis, venis oblique patentibus, costis subtus deciduo-squamulosis, fertilibus paulo minoribus subintegerrimis vel sinuato-lobulatis dorso (ut videtur) toto capsuliferis.

Acrostichum dimorphum, *Hook. et Grev. Ic. Fil. Tab.* 145, *frons fertilis inclusa*, (not "*exclusa*," Fée). *Fée, Hist. des Acrost.* p. 40.

Olfersia dimorpha, *Pr. Tent. Pterid.* p. 235.

Elaphoglossum dimorphum, *Moore.*

HAB. St. Helena, on rocks and walls, *Genl. Walker, Dr. Shuter* (who also sent specimens from Madras, but they were probably taken out thither from St. Helena); top of Diana's Peak, very common, *Dr. Hooker.*

This and the subject of our following plate, *Acrost.* (Elaphoglossum) *bifurcatum* are very remarkable plants; peculiar to the island of St. Helena; and, as Dr. Hooker assures us, growing apart from each other; and, where he has seen them, exhibiting no appearance of being forms of one and the same species. Nevertheless, M. Fée, who does not admit, in his Histoire des Acrostichées, the *A. furcatum*, observes at p. 40. under *Acrostichum dimorphum*, Hook., "cette espèce est parfaitement tranchée, et la dissimilitude des feuilles ne permettra pas de la méconnaitre." He goes on to say "*l'A bifurcatum* pourrait bien n' être autre chose q'une forme tres-divisée de *l'A. dimorphum.*" Presl, having in the interim established his genus *Microstaphyla*, with no character except form, that I can see, to distinguish it from *Elaphoglossum* (his *Olfersia*); ("sorus," he says "acrostichoideus"), founded on a solitary species, the *Acrost. bifurcatum*, Sw., M. Fée in his 7me "Mém. sur les Fougéres," adopts the Genus, repeats his views respecting the oneness of *Acr. dimorphum* and *Acr. bifurcatum*, and asserts that our figure of the former "pêche par l'exactitude, la fronde fertile figurèe n'appertenant pas vraisemblablement aux frondes steriles." He describes and represents the sori of *Microstaphyla* as "*nervillaires*," which we cannot confirm by our specimens. We must refer our readers to the next following description for further observations on these two Ferns.

TAB. XC. Fertile and sterile fronds of *Acrostichum (Elaphoglossum) dimorphum*, Hook. and Grev.; *natural size. Fig.* 1. Portion of a sterile, and *f.* 2. portion of a fertile frond; *magnified.*

Tab. XC

TAB. XCI.

ACROSTICHUM (ELAPHOGLOSSUM) BIFURCATUM, *Sw.*

Glaberrimum nudum, caudice horizontali vel ascendente crassiusculo radicante apice solummodo squamoso, stipitibus
dense cæspitosis gracilibus 3-6-uncialibus stramineo-fuscis,
frondibus 2-4-uncialibus oblongo-lanceolatis pinnatis, rachibus alatis : pinnis *sterilium* linearibus remotis simplicibus
plerisque furcatis vel bifurcatis costatis seu uninerviis, *fertilium* pinnis approximatis brevioribus cuneatis vel subquad-
ratis apice bi-4-fidis vel bi-trifurcatis.

Acrostichum bifurcatum, *Sw. Syn. Fil.* p. 42. *Schk. Fil.* t. 3.
Willd. Sp. Pl. 5, p. 114.

Osmunda bifurcata, *Jacq. Col.* t. 20, f. 2.

Olfersia bifurcata, *Pr. Tent. Pterid.* p. 234.

Darea furcans, *Bory, Voy. de la Coquille, Bot.* p. 269, t. 35,
f. 2. *(sterile).*

Anogramme parodoxa, *Fée, Gen. Fil.* p. 64.

Gymnogramme bifurcata, *Kze in Linnæa,* 10, p. 496.

Microstaphyla furcata, *Pr. Epimel. Bot.* p. 161. *Fée,* 7me
Mém. des. Foug. p. 45.

Polybotrya bifurcata, *Moore.*

HAB. St. Helena, and judging from the quantity of specimens we
have received from different voyagers, it must be infinitely more
abundant than the subject of our last plate, *Acr. dimorphum :*
on wet rocks and mossy banks, to an altitude of 1000 feet and
more above the level of the sea *(Hook. fil.). Plukenet* records it and
figures it more than 160 years ago as " Filicula corniculata *Insulæ
Sanctæ Helenæ,*" &c. My Herbarium contains specimens from
the late *Sir G. Staunton,* collected on the voyage of Lord Mac-
artney's Embassy to China, from *Menzies, Dr. Shuter, Cuming,*
n. 420 and 421, *Nuttall, Lady Dalhousie, Dr. Lyall, Seeman, J. D.
Hooker,* but it is only the latter and Dr. Shuter who appear to
have gathered the *A. dimorphum ;* also *Forster, Lichtenstein,* and
Liebold are recorded as having gathered this species.

I have already stated under *Acrost. dimorphum* (see our last
plate XC.) that M. Fée pronounces that and the present plant to
be one and the same, without even making a variety. " Nous
avons sous les yeux diverses modifications qui semblent établir
le passage de l'une à l'autre par des nuances insensibles." My
own copious specimens I must confess lead me to an opposite con-
clusion, and I have represented in our figures its extreme forms.
A. bifurcatum has much longer and slenderer stipites, always destitute of scales, as is every part of the plant. It is true that in the
fertile fronds there is a great tendency to become entire, less deeply
divided than in the sterile fronds (as is still more strikingly seen in
an allied group of *Acrostichum, Rhipidopteris,* Schott), but it goes
no further than our figures show. If *A. dimorphum* were to break
up into *A. bifurcatum,* the venation must undergo a considerable
change ; for whereas in *A. dimorphum* the veins are simple or only
once forked, here they are not uncommonly twice forked, the segments having the same ramification ; hence the specific name.

TAB. XCI. Sterile and fertile fronds of *Acrostichum (Elaphoglossum) bifurcatum,* Sw. ; *natural size. Fig.* 1. Pinna of a sterile frond,
and *f.* 2 and 3, Pinna of fertile fronds ; *magnified.*

Tab. XCI.

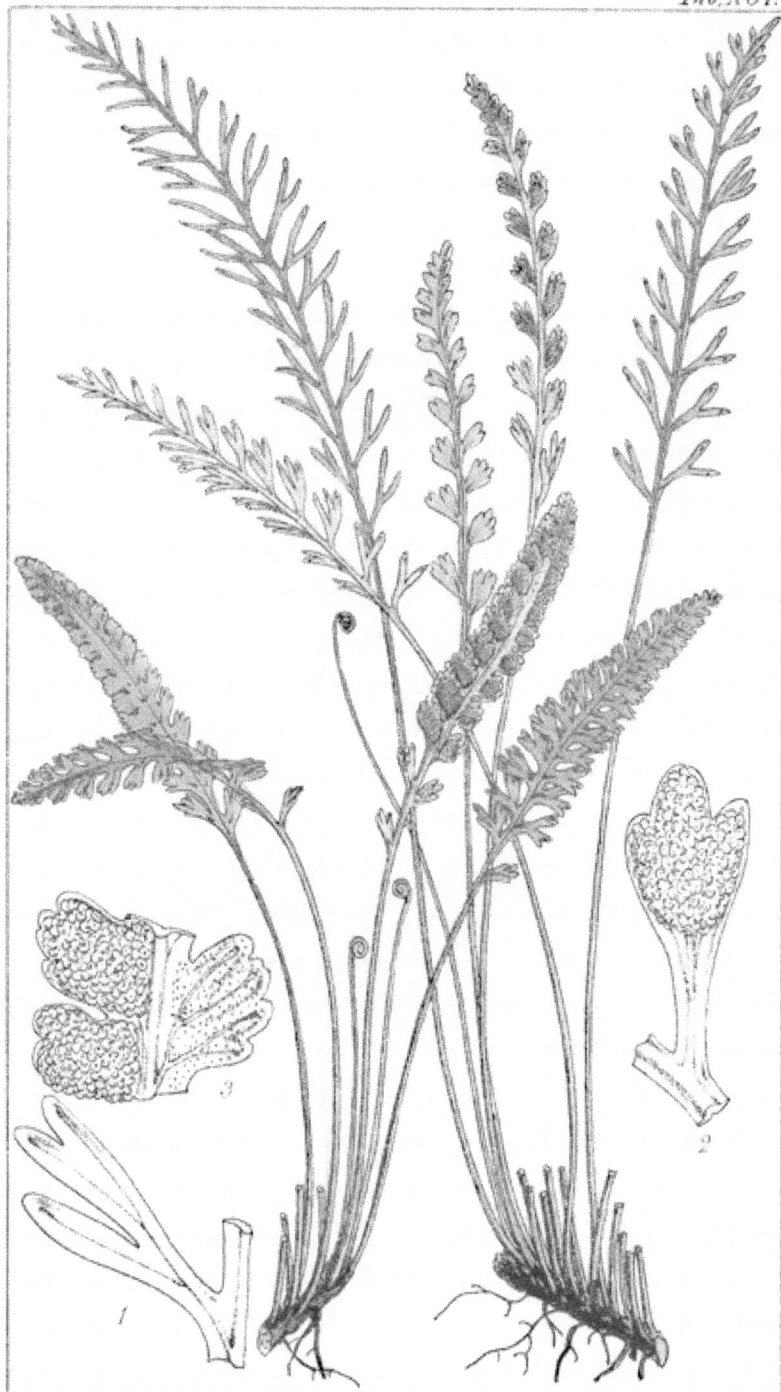

TAB. XCII.

ACROSTICHUM (ELAPHOGLOSSUM) FÉEI, *Bory.*

Caudice longe repente et, ut videtur, arboribus scandente
radicoso squamis ovato-acuminatis paleacis, stipitibus sparsis
paleaceis; frondibus *sterilibus* sesquiuncialibus oblongo-lanceolatis obtusis grosse subcrenato-lobatis parce squamulosis
in petiolum æquilongum basi attenuatis costatis pinnatim
venosis, venis remotis infra medium furcatis ante marginem
terminantibus apicibus clavatis; *fertilibus* subduplo minoribus elliptico-oblongis integerrimis petiolo duplo triplove
brevioribus.

Acrostichum Féei, *Bory, in Fée, Acrostich.* p. 48, tab. 18, f. 2.
Elaphoglossum Féei, *Moore*

HAB. Tropical America, Guadeloupe "*de Thiouville,*" *L'Her-
minier, in Herb. Nostr.* On Mount Couliabon, Dominica,
Dr. Imray. Quito, *Jameson.*

This is justly considered, by M. Fée, who has first described and figured this pretty *Acrostichum*, to have the habit
of *Polypodium serpens,* Sw.

TAB. XCII. Plant of *Acrostichum (Elaphoglossum) Féei,*
Bory; *natural size. Fig.* 1. Sterile frond; *f.* 2. Portion of
a sterile frond, and *f.* 3, portion of a fertile frond; all *more or
less magnified.*

Tab. XCII

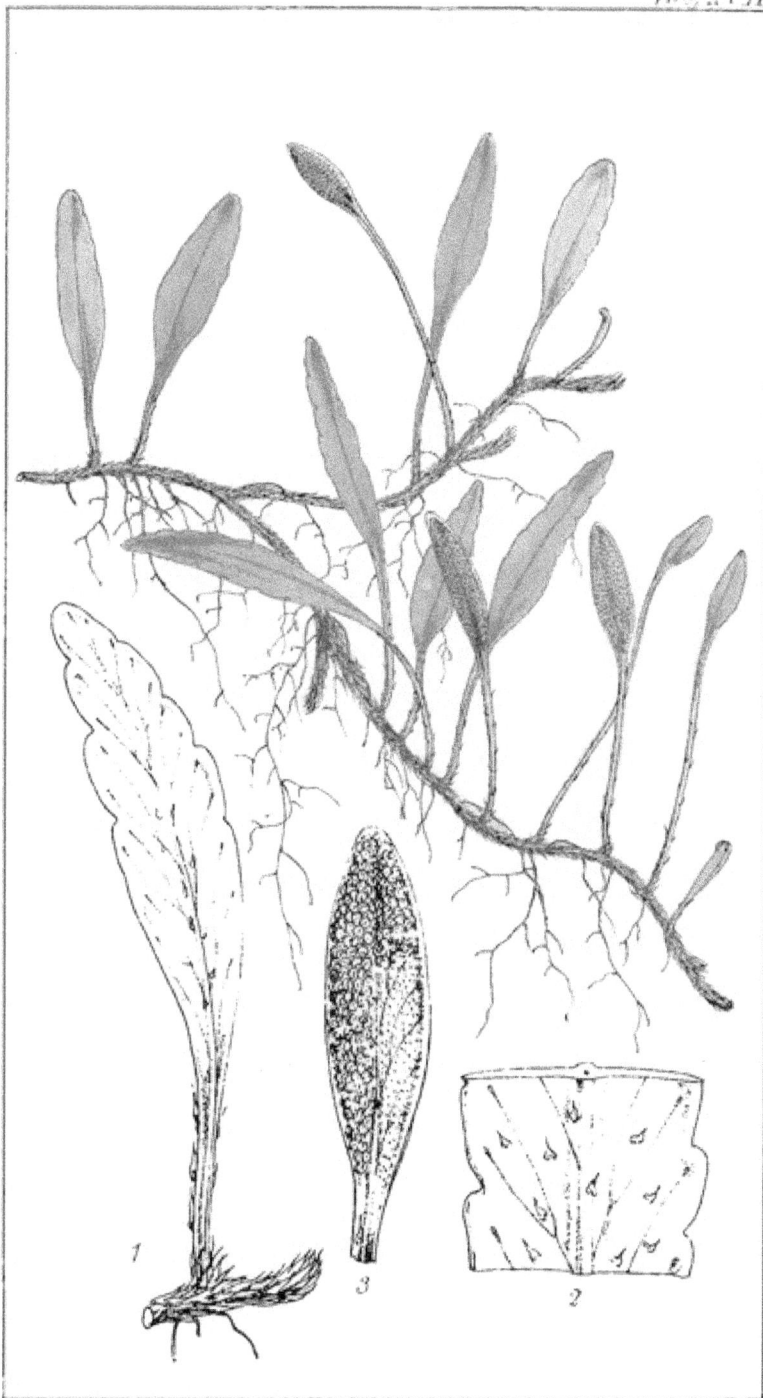

TAB. XCIII.

DAVALLIA (DAREOIDEÆ) NIGRESCENS, *Hook.*

Stipite pedali et ultra subrobusto viridi-fusco squamis subulatis parvis paleaceo, fronde sesquipedali ad bipedalem late
ovata acuminata submembranacea flaccida glaberrima siccitate nigrescente 3-4-pinnata, pinnis primariis secundariisque ovato-lanceolatis longe acuminatis petiolatis ultimis
pinnatifidis, laciniis oblongis ovatisve subsecundis acutis
integerrimis uninerviis, nervis infra apicem terminantibus,
soris ad marginem superiorem pinnularum seu laciniarum
insertis prominentibus, involucro calyciformi (hemispherico)
firmo frondis colore margine integerrimo, rachibus ubique
planiusculis læviter marginato-alatis.

HAB. Peak of Fernando Po; elev. 3000 feet. *Gustav. Mann,*
n, 448.

It is difficult to say whether this should be referred to
Davallia or to *Dicksonia.* If considered a *Davallia,* the portion of the frond to which the involucre is attached, is so
tumid as to represent, on the upper side of the frond, the involucre on the underside, thus exhibiting a perfectly hemispherical cup with an entire margin, but sunk in the frond. In
habit and ramification it very much resembles *Asplenium*
(Darea) *Shuttleworthii* of Kunze in Schk. Fil. Suppl. p. 26,
t. 14; but there the sorus and involucre are truly dareoid,
laterally elongated; here hemispherical.

TAB. XCIII. *Fig.* 1. Portion of stipes of *Davallia nigres-*
cens, Hook.; and *f.* 2 and 3, apex and primary pinna of a
frond; *natural size. f.* 4. Ultimate fertile pinna with sorus;
magnified. f. 5. Sorus from which the involucre is removed,
showing the long stipitate capsules; *more magnified.*

Tab. XCIII

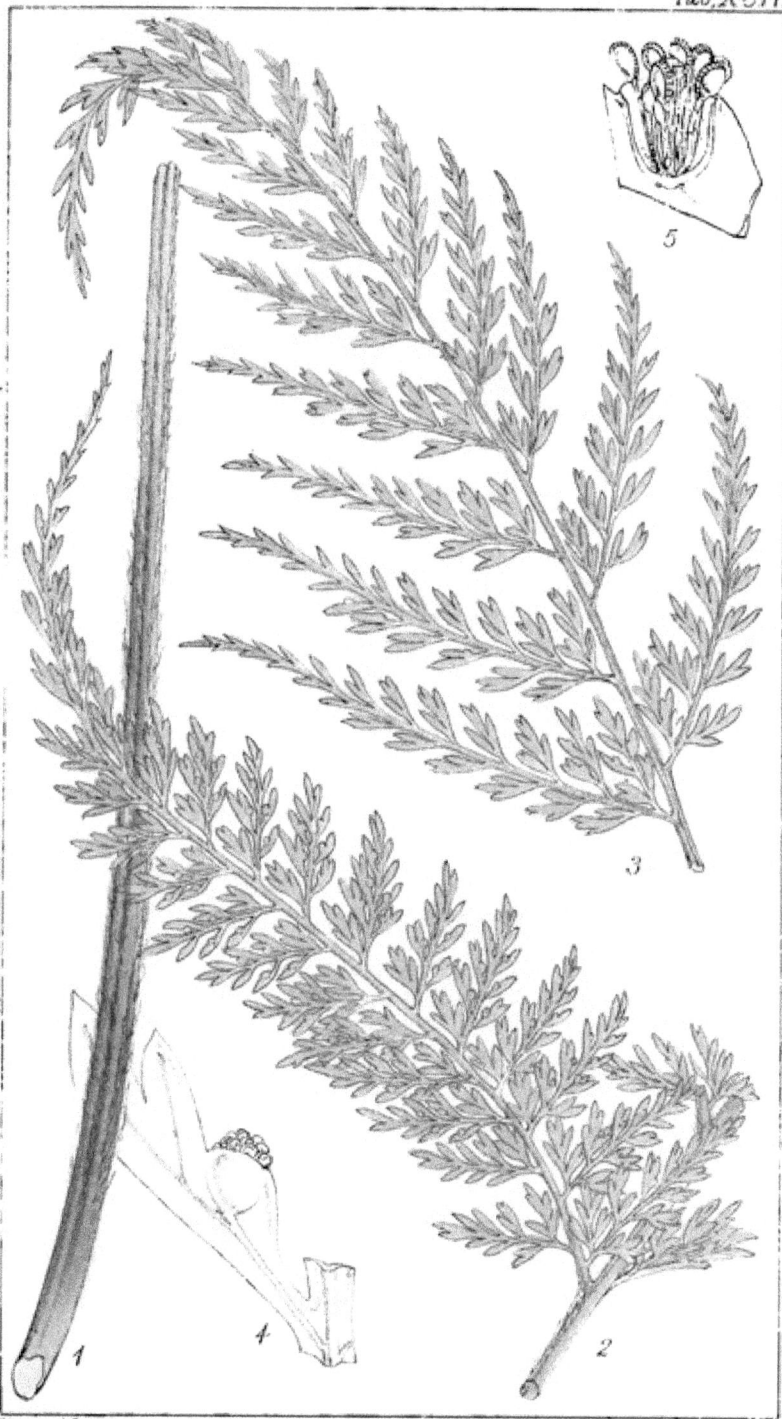

TAB. XCIV.

HELMINTHOSTACHYS ZEYLANICA, *Hook.*

Gen. Char. HELMINTHOSTACHYS, Kaulf. *Caupsulæ* subglobosæ, glomeratæ, in valvas duas subæquales deorsum longitudinaliter dehiscentes: *glomeruli* pedicellati apice cristati, in spicam elongatam pedunculatam cordiformem dense congesti. *Sporæ* hyalinæ parvæ subglobosæ.—Rhizoma *crassum, carnosum, repens, fibrosum,* fibris *carnosis.* Stipes *simplex, elongatus, apice frondosus,* fronde *ternata ad basin pedunculi verticillata,* foliolis *petiolatis pinnatis, pinnis subquinis lanceolatis, maqis minusve acuminatis, minute serratis, lateralibus longe decurrentibus,* terminali *libera.*

Helminthostachys Zeylanica.

Helminthostachys Zeylanica, *Hook. Gen. Fil.* t. 47, *Presl, Suppl. Tent. Pterid.* p. 59.

Helminthostachys dulcis, *Kaulf. En. Fil.* 28, t. 1, f. 1 *(spike). Blume, En. Fil. Jav.* p. 258, *Wall. Cat. n.* 54, *Hook. et Grev. En. Fil. in Bot. Miscell.* 3, p. 220.

Botrychium Zeylanicum, *Sw. Syn. Fil.* p. 172, *Willd. Sp. Pl.* 5, p. 65.

Osmunda Zeylanica, *Linn. Sp. Pl.* p. 1519.

Ophioglossum laciniatum, *Rumph. Herb. Amb.* 6, p. 153, t. 68, f. 3.

Hymenostachys integrifolia, *Presl, Suppl. Tent. Pterid.* p. 50.

Botryopteris Mexicana, *Presl, Rel. Hænk.* 1, p. 76, t. 12, f. 1.

Hymenostachys crenata, *Presl, Suppl. Tent. Pterid.* p. 60.

HAB. Malayan and Molucca islands, frequent; Amboyna, *Rumph;* Luzon, *Cuming, n.* 39; Tavoy and Bengal, *Wallich;* Mergui, *Griffith;* Java, *Blume, Thos. Lobb;* Borneo, *Thos. Lobb;* Ceylon, *Gardner;* Cochin, *Mrs. General Walker, Johnston;* New Caledonia, *La Billardiere;* Guahan, Marianne Islands, *Hænke.*

Of this remarkable Genus there is but a solitary species known: notwithstanding that Presl makes three species out of that distributed by Cuming from Luzon. One of these is the same as the *Botryopsis Mexicana* of the same author in the "Reliquiæ Hænheanæ;"—afterwards, on finding that it was not a Mexican plant, and a *Hymenostachys,* the name was changed to *H. crenata.* The figure is a very good representation of the true *H. Zeylanica,* save the reticulated venation, which is no doubt occasioned by the ingenuity of the artist.

TAB. XCIV. Fertile plant of *Helminthostachys Zeylanica,* Hook.; *natural size. Fig.* 1. Portion of a sterile frond, showing the venation. *f.* 2. Portion of a fertile spike, with clusters of capsules; and *f.* 3. single cluster of capsules with its crest; all *more or less magnified.*

Tab. XCIV

TAB. XCV.

MARATTIA (EUPODIUM) KAULFUSSII, *J. Sm.*

Fronde " vasta " tri-quadripinnata, pinnulis ultimis oblongis ovalibusve pinnatifidis obtusis acutisve pinnatis, venis simplicibus vel furcatis, rachibus junioribus late interrupte alatis, soris in quoque lobo solitariis v. binis, involucro pedicellato.

Marattia Kaulfussii, *J. Sm. in Hook. Gen Fil. sub. tab* 26 (note). *Metten. Fil. Hort. Lips.* p. 118. J. G. Sturm, *in Mart Fil. Bras.* p. 153.

Marattia lævis, *Kaulf. Enum. Fil.* p. 31 (excl. syn. omn.)

Marattia alata, *Raddi, Fil. Bras.* t. 83-84, (excl. syn. omn.)

Eupodium Kaulfussii, *J. Sm. in Hook. Journ. Bot.* 4, p. 190, *and in Hook. Gen. Fil.* t. 118. *Presl, Suppl. Tent. Pterid.* p. 17, *De Vriese, Monogr. Marutt.* p. 12. *Brackenr. Fil. U. S. Exp. Exped.* p. 313.

HAB. Tropical S. America: Brasil, about Rio, *Raddi, Beyrich, Dr. Lyall, J. D. Hooker, Martius, Vautier, Riedel* and others. Antioquia, *Jervise*; Columbia, *Triana, Purdie*; Venezuela, *Fendler,* n. 3 ; Caraccas, *Linden,* n. 196.

I fear both the genera and species of the *Marattiaceous* Ferns have been needlessly multiplied. The present one, however, is well distinguished from both *Marattia lævis* and *M. alata,* with which it had been confounded, and as clearly shown by Mr. J. Smith, by the presence of a distinct pedicel to the sorus. There is a further character in the uniformly pinnatifid ultimate pinnæ : but the entire genus requires to be carefully elaborated. I cannot agree with those who consider this a distinct Genus, on account of the pedicellated fruit.

TAB. XCV. *Fig* 1. Inferior portion of a fertile or primary pinna; and *f*. 2. apex of the same, of *Marattia (Eupodium) Kaulfussii,* J. Sm. *natural size. f.* 3. Ultimate sterile pinna and portion of the rachis, and *f*. 4. portion of a fertile pinna, with sori; *magnified. f.* 5. Sorus, with its pedicel, *more magnified.*

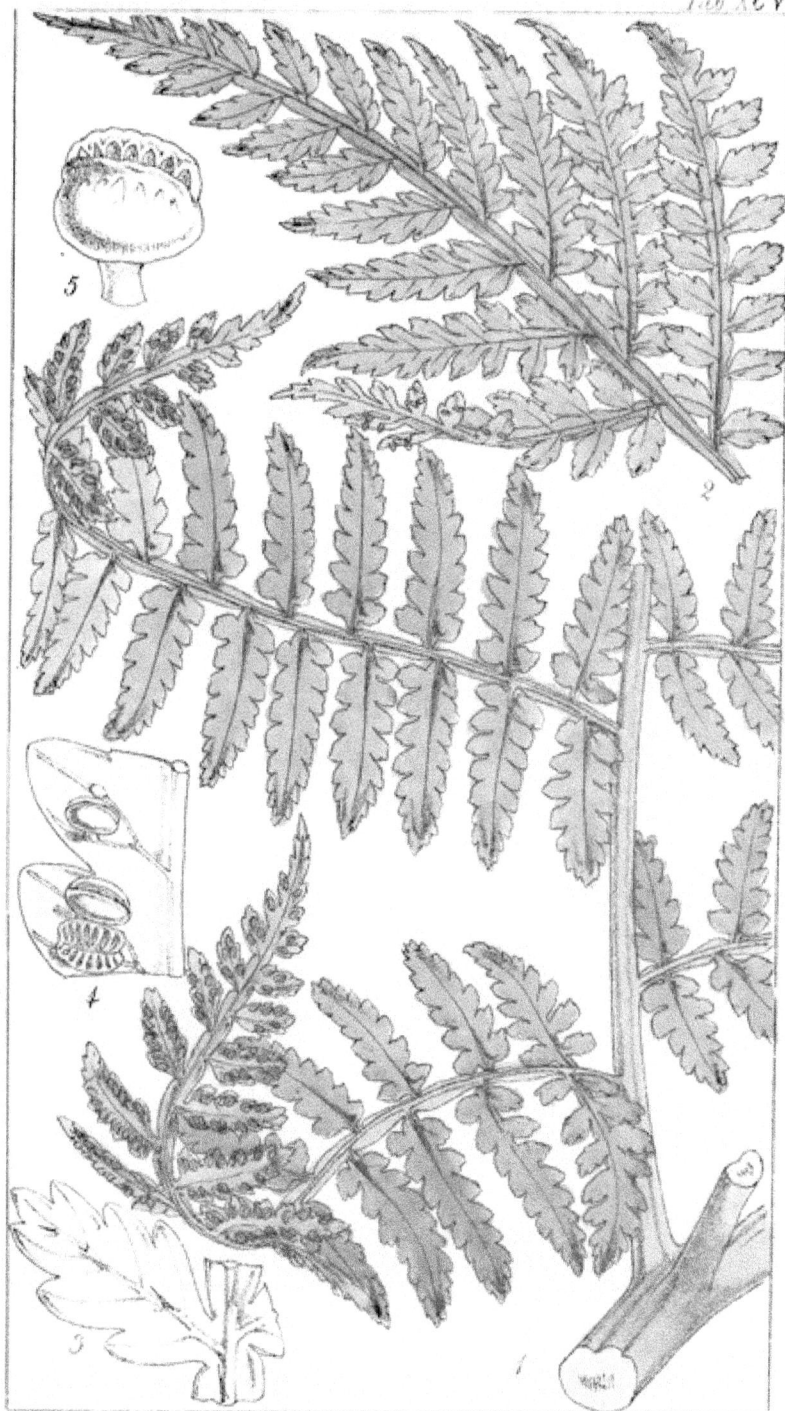

Tab XCV

TAB. XCVI.

DAVALLIA (LEUCOSTEGIA) PILOSELLA, *Hook.*

Laxe molliter villosa, caudice repente apice paleaceo, stipitibus gracilibus sparsis patenti-villosis 4-5 uncias longis, frondibus membranaceis oblongis acuminatis, junioribus pinnatis,
  pinnis oblongis obtusis serrato-lobatis, adultis subspithamæis
  pinnatis, pinnis petiolatis ovato-lanceolatis acuminatis profunde pinnatifidis, laciniis oblongo-lanceolatis acutis serrato-
  pinnatifidis, soris in dentibus vel lobulis laciniarum cuneato-
  rodundatis longe villosis, venis pinnatis.

HAB. Tsus Sima, island off the coast of Corea, *Wilford,* n. 793.

A very delicate *Davallia,* which I refer to the *Leucostegia-*
group, and which may rank near to *Davallia membranacea,*
Wall.; differing in the more contracted frond, the differently
shaped sori, which are here quite marginal on short teeth or
lobes, but especially in the copious long, soft, white, jointed
hairs of the entire frond.

TAB. CXVI. Plant with fertile and sterile frond of *Davallia pilosella,* Hook.; *natural size. Fig.* 1. Segment of a
frond with two sori; *magnified. f.* 2. single sorus, *more magnified*; and *f.* 3. single sorus, with the involucre removed,
showing the capsules, *ditto.*

Tab. XCVI

TAB. XCVII.

POLYBOTRYA LECHLERIANA, *Metten.*

Caudice crasso elongato scandente squamis ovatis acuminatissimis fuscis paleaceo, stipitibus fuscis nitidis approximatis spithamæis 2-3 lineas latis inferne squamosis; fronde *sterili* ampla 3-pedali et ultra submembranacea firma ubique (rachibusque squamulosis testaceis) villosa late ovata acuminata tripinnata, pinnis primariis subsessilibus spithamæis ad pedalem remotis ovatis acuminatis, secundariis approximatis numerosis, pinnulis ½-unciam longis copiosis ovato-oblongis profunde pinnatifidis segmentis oblongis acutiusculis vix subfalcatis univeniis, venis apice clavatis ; *fertili* pinnulis linearibus obtusissimis lobato-pinnatifidis subtus utrinque fere ad marginem soriferis.

Polybotrya Lechleriana, *Metten. in Hohenack. et Lechl. Plantæ Peruvianæ,* n. 2156, *et Fil. Lechl. Chil et Peru,* p. 4, t. 1, ff. 1-5.

HAB. Shady places, St. Gavan, Peru, *Lechler;* Mount Guayrapurime, near Tarapota, eastern Peru, sterile frond only, sent by *Mr. Spruce,* 1856. Ecuador, sent by *Dr. Jameson,* exact locality not stated.

This is the finest and most copiously divided of any of the Acrostichaceous group of Ferns. Fragments only were known to its describer, Dr. Mettenius: but a magnificent frond has come into my possession along with Lechler's own collection of Ferns of Peru and Chili. The only other specimens known to me are a fine barren frond with caudex, gathered at Tarapota by Mr. Spruce, and the lower half of a fertile frond sent from Ecuador by Professor Jameson. The form of the sterile pinnæ and segments a good deal resembles some *Darea* among *Asplenium,* or our *Davallia nigrescens,* represented at our plate XCIII. It will rank near *Polybotrya apiifolia.*

TAB. XCVII. Caudex and base of a sterile frond, and primary pinna of a fertile frond of *Polybotrya Lechleriana.* Metten.; *natural size. Fig.* 1. Pinnule of a sterile frond; *f.* 2. Fertile pinna seen from above, and *f.* 3. fertile pinna seen from beneath ; *magnified. f.* 4. Transverse section of a fertile segment ; *more magnified.*

Tab. XCVII

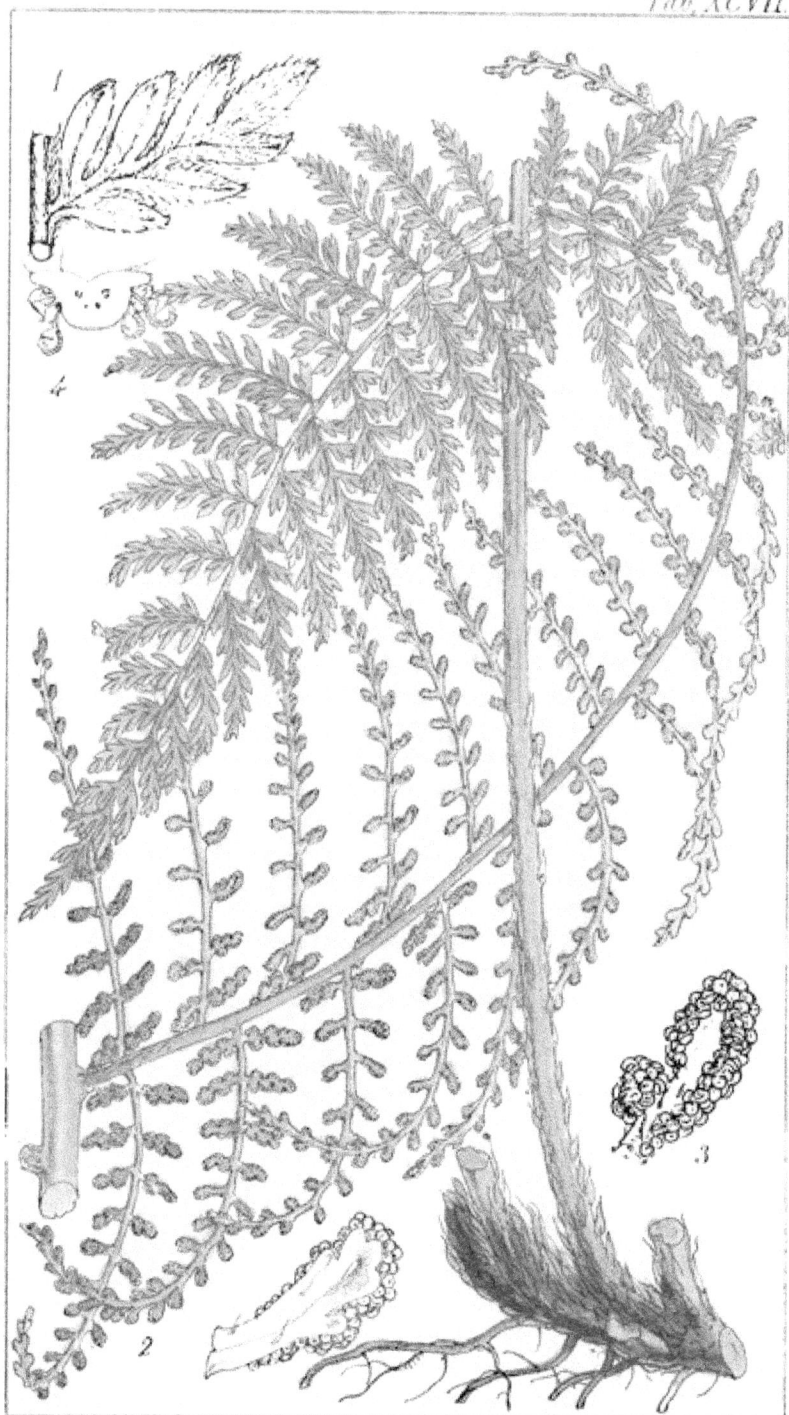

TAB. XCVIII.

WOODSIA (PHYSEMATIUM) MANCHURIENSIS, *Hook.*

Glaberrima, caudice perbrevi fibroso ferungineo-squamoso, stipitibus cæspitosis 1-3-uncialibus gracilibus rachibusque albo-stramineis nitidis, frondibus spithamæis tenui-membranaceis oblongo-lanceolatis acuminatis pinnatis, pinnis sessilibus unciam et ultra longis remotiusculis oblongis obtusis pinnatifidis, lobis oblique patentibus ovatis lato-oblongisve integris vel sinuato-sublobatis unisoris, venis in singulo lobo pinnatis, venulis apice clavatis venula infima superiore ante apicem sorifera, involucro (ratione plantæ) majusculo tenui-membranaceo globoso demum apice irregulariter rupto.

HAB. Manchuria, *C. Wilford,* 1859, n. 1094.

This new species of the *Physematium*-group of *Woodsia,* quite distinct from any hitherto described, was detected by Mr. Wilford, in a country whose Botany is deserving of a careful investigation, Manchuria. The involucres are of so delicate a texture that the pressure given in drying destroys their spherical form, and in that state they present the appearance of a thin membrane, but soaking in warm water restores them to their proper shape.

TAB. XCVIII. Plant with fertile fronds of *Woodsia (Physematium) Manchuriensis,* Hook.; *natural size. Fig.* 1. Fertile pinna with sori ; *magnified. f.* 2. Segment of a fertile lobe showing the venation and sorus; *f.* 3. the same with the sorus (*f.* 4.) removed from the lowest superior vein; and *f.* 5. Old involucre, burst open; *more highly magnified.*

Tab. XCVIII.

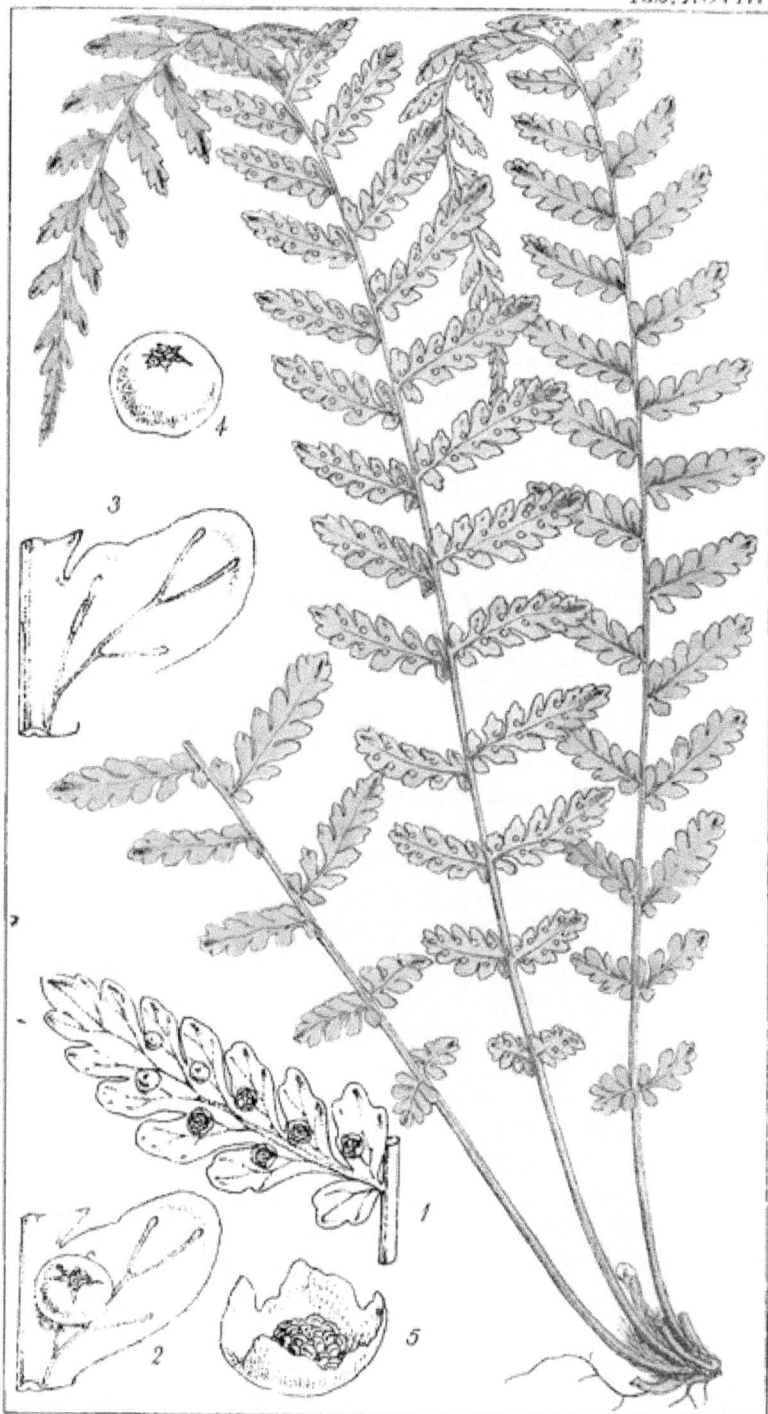

TAB. XCIX.

Cyathea microphylla, *Metten.*

Caudice 4-pedali (Lechl.) frondibus 2-3-pedalibus stipitatis
(stipitibus rachibusque decidue ferrugineo-tomentosis) oblongo-ovatis acuminatis coriaceo-membranaceis tripinnatis
siccitate fusco-olivaceis, pinnis primariis sessilibus horizontalibus remotiusculis oblongis acuminatis, secundariis approximatis oblongis obtusis, pinnulis ovato-oblongis profunde
pinnatifidis, lobis oblongis subcuneatisve obtusissimis integerrimis univeniis, venis ante apicem evanidis, soris copiosis
dorso venarum basin versus globosis membranaceis ferrugineis nitidis demum apice irregulater fissis et in lacinias
3-4 ruptis, receptaculo elevato, capsulis helicogyratis, rachibus ultimis squamis paucis convexis flexuosis ferrugineis
deciduis subtus paleaceis.

Cyathea microphylla, *Metten. in Hohenack. et Lechl. Plant.
Peruv. n.* 2569, *and in Fil. Lechl. Chil. et Peruv. p.* 23.

Hab. Tatanara, eastern side of the Cordilera of Peru, *Lechler, in Herb. Nostr. n.* 2160.

A very remarkable Cyatheaceous plant, the most distinct
perhaps of all the group. The underside of the fertile frond
has quite a ferrugineous cast, from the hairs of the tomentum
of that color, and from the copious capsules, which soon,
from their original spherical form, break into a number of
pieces or scales, and in a measure conceal the real shape of
the small pinnules.

Tab. XCIX. *Fig.* 1. Base of a sterile frond; and *f.* 2.
Underside of a primary pinna of a fertile frond of *Cyathea
microphylla.* Metten. *natural size. f.* 3. Fertile pinnule with
two sori, *magnified.* and *f.* 4. fully formed sorus, and *f.* 5.
old sorus; *more magnified.*

Tab. XCIX

TAB. C.

HEMITELIA (AMPHICOSMIA) PLATYLEPIS, *Hook.*

Stipite elongato castaneo nitido versus basin crassitie digitis
humani, squamis maximis ovatis nitidissimis acuminatis
atro-fuscis margine pallidioribus suberosis basi squamulosis
paleaceis, frondibus amplissimis subcoriaceis ubique hirsu-
tulis tripinnatis, pinnis primariis inferioribus longe petiolatis
pedalibus sesquipedalibus ovato-oblongis, secundariis sessilibus oblongo-lanceolatis acuminatis, pinnulis 3-4 lineas
longis linearibus acutis marginibus recurvatis serratis, venis
furcatis ad furcaturam unisoris, rachibus adpresse villosis
ultimis alatis, involucro exacte hemisphærico membranaceo
margine ereoso-denticulato, receptaculo elevato, capsulis
pilis articulatis intermixtis.

HAB. Near San Carlos, Rio Negro, tributary of the Amazon,
Brazil, *R. Spruce, n.* 3027.

This is a *Hemitelia*, according to our views of the Genus,
but would be an *Amphicosmia* of the late Mr. Gardner, Lond.
Journ. of Bot. 1, p. 441; differing from *Hemitelia* in its free
venation. The involucre is exactly the same as in our *H. Host-
manni* and *H. Parkeri*, described in the "Species Filicum," and
figured in the 7th volume of our "Icones Plantarum." It is
probably nearly allied to *Cyathea multiflora* of Sir J. E. Smith,
which has the "rachis winged," and which Gardner refers to
Amphicosmia: but I have never seen an authentic specimen,
which I believe only exists in the Banksian Herbarium, and
which is stated to be a native of Jamaica. In general the
specimens we receive of these gigantic Ferns are too imperfect
for accurate description, even the stipes is often neglected,
which not unfrequently affords good characters in the pecu-
liarities of its clothing especially, as is remarkably the case
in the species now under consideration; and, the involucre
being very fragile, is too frequently injured, and leads one
astray in regard to the Genus.

TAB. C. *Fig.* 1. Base of the stipes of *Hemitelia (Amphicosmia) platylepis*, Hook., with its remarkable large scales
and the squamules at their point of insertion; and *f.* 2.
portions of a primary pinna of a fertile frond; *natural size.*
f. 3. Under side of a barren pinnule, with a portion of the
winged rachis; *magnified. f.* 5. Portion of a fertile pinnule,
with a sorus; *f.* 6. Involucre and receptacle from which
most of the capsules have fallen; and *f.* 7. capsules with
accompanying hairs; *more magnified.*

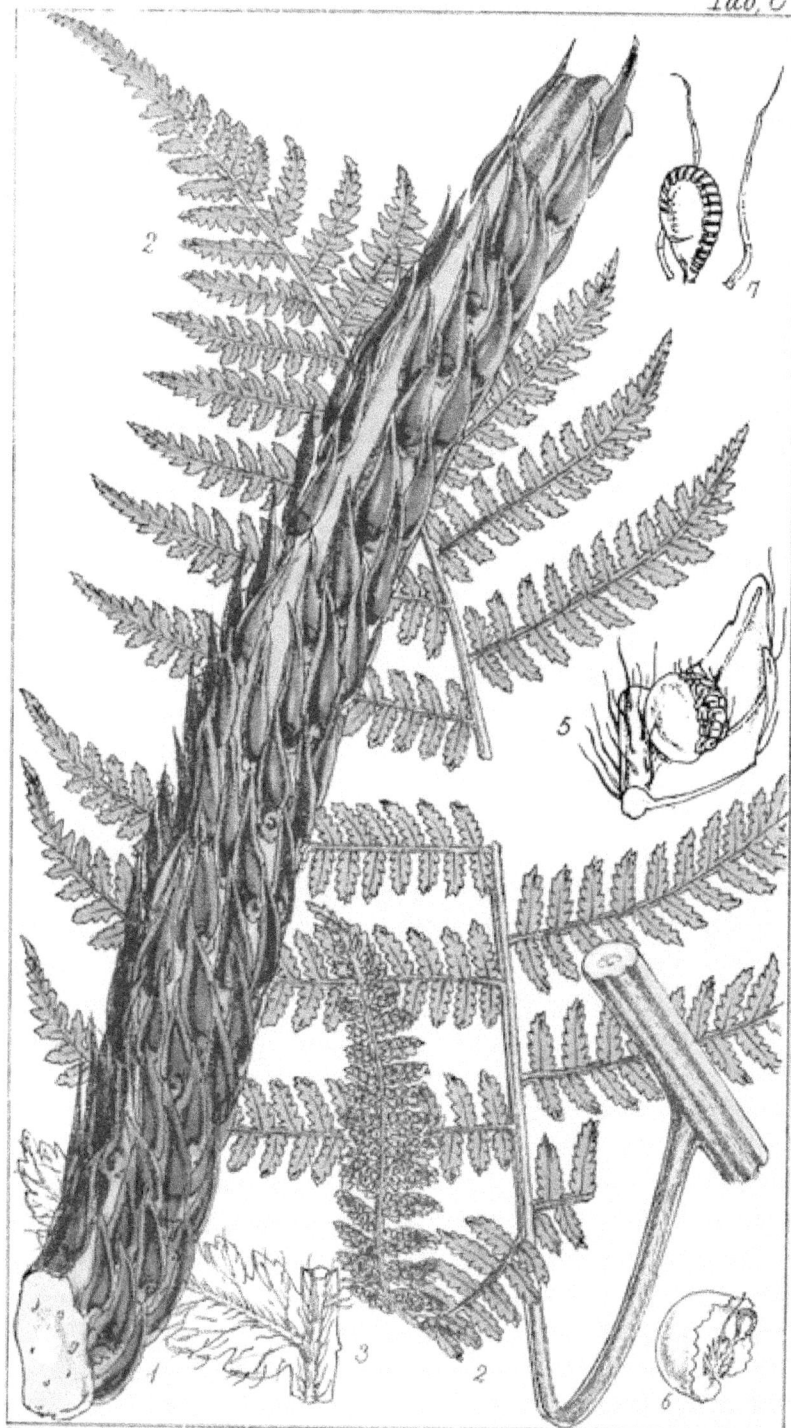

Tab. C

www.ingramcontent.com/pod-product-compliance
Lightning Source LLC
Chambersburg PA
CBHW021349210326
41599CB00011B/809